Objective
Based
Selling™

How to sell more material handling equipment (by focusing on the customer instead of the stuff!)

Gary T. Moore

Trent Press

DENVER

For information, contact
Trent Press
556 Franklin St.
Denver, CO 80218
www.trentpress.com

Objective Based Selling™
LCCN: 2007906492
ISBN-10: 0-9799241-0-3
ISBN-13: 978-0-9799241-0-1

Production Management by
Paros Press
1551 Larimer Street, Suite 1301   Denver, CO 80202
303-893-3332   www.parospress.com

Book Design by Scott Johnson

Printed in the United States of America
1   3   5   7   9   10   8   6   4   2

Table of Contents

Material Handling Industry Overview

In his book *The World Is Flat,* author Thomas L. Friedman lists ten trends that have "flattened" the world, reducing barriers to information exchange, trade, and globalization. Number seven on his list is logistical supply chaining.

Material handling is the industry that has evolved into the logistical supply chain. Material handling encompasses the equipment, technology, methods and services to move, store, identify, and deliver materials around the world.

Three industry association websites give excellent overviews of the industry:

Material Handling Equipment Distributors Association:
        www.mheda.org
Material Handling Industry of America: www.mhia.org
Industrial Truck Association: www.indtrk.org

Each of these sites provides an introduction to the industry and to the types of equipment and services used and sold as part of the material handling industry and logistical supply chain. There are many links from these sites to the websites of individual companies in the industry.

Material handling is essential to the functioning of society and is certainly critical to the competitiveness of individual companies and countries competing in the global marketplace.

Most of the equipment and services in this industry is sold business-to-business, by individual salespeople. This book introduces a sales model to help salespeople of material handling equipment and services sell more at higher gross margins. In certain parts of the book, words denoting specific types of equipment will be used as examples. These are not meant to be limiting. **Objective Based Selling,** properly executed, can be used to effectively sell the entire range of material handling products and services.

Sales Goes to Hollywood

In order to approach selling as a profession, and excel, many salespeople must overcome negative societal stereotypes of salespeople and selling. These stereotypes are regularly reinforced in Hollywood movies.

A brief history of salespeople (in the case of Hollywood, more accurately labeled sales*men)* in movies includes:

"Death of a Salesman" (1951) based on the play by Arthur Miller

Willy Loman, the traveling salesman of the title, is a failure—at selling and at life. His apparent primary sales skill has been "glad handing." Early in the movie he is fired for not achieving. He also is revealed to have cheated on his wife. He loses the respect of one of his sons. He commits suicide. Willy is probably the most famous fictional salesman. Great role model.

"Tin Men" (1987)

Bill "BB" Babowsky is one of two fast-talking aluminum siding salesmen (tin men) in this story set in the 1960's. They talk their way into homes and trick unsuspecting homeowners into signing contracts for siding they probably don't need and can't afford. A favorite trick is to surreptitiously drop a five-dollar bill on the floor and "return" it to the housewife as a sign of honesty. A great example of rapport-building.

"Cadillac Man" (1990)

Joey O'Brien (played by Robin Williams) is the stereotypical fast-talking (notice how often this trait appears in the depiction of salesmen) sleazy car salesman. Enough said.

"Glengarry Glen Ross" (1992)

Jack Lemmon plays Shelly Levine, a fading real estate sales-man, "smiling and dialing" older couples. Shelly is desperate to meet his quota and save his job, as well as get money for a daughter's medical procedure. Alec Baldwin personifies the high-pressure sales manager conducting a sales contest. First prize: new Cadillac. Second prize: set of steak knives. Everybody else gets fired. Baldwin gives a motivational speech involving prominent display of a large pair of brass balls.

"Fargo" (1996)

In this dark comedy (in which six people are murdered), a dis-honest car salesman (now, there's a novel concept) hires two hap-less losers to kidnap his wife, in order to extort ransom from her rich father. On the job, he finances cars he doesn't own, and cheats customers. Instead of holding his wife for ransom, the kidnappers kill her. The most enduring image of the movie is one of the killers being fed headfirst into a wood chipper. Makes you want to be a salesman, doesn't it?

"Boiler Room" (2000)

Seth Adams is every telemarketer's role model—and worst nightmare for every couple quietly eating dinner. Seth sells worthless stock over the phone—pumping up the value so his bosses can dump the stock (hence, the endearing sales term "pump and dump"). About to be caught, Seth finally cooperates with the authorities. Our hero.

A search for countervailing, positive images of salespeople on the silver screen resulted in... none.

These negative stories and depictions of salespeople were not created by Hollywood. Unfortunately, they are reflections of the perceptions of much of the general public.

Yet selling remains one of the highest-paid, most critical professions of free enterprise economies. It's one of the jobs that rewards performance (or lack of it) relatively proportionate to results produced. Selling is the catalyst that makes things happen, putting millions of people to work manufacturing products, performing services and filling orders.

The old business cliché is a cliché because it is true:

"Nothing happens until somebody sells something."

For 80 years, an economic experiment was tried without salespeople. It was called communism.

If selling is so important, why are sales stereotypes so consistently negative?

Two reasons:

- Most people—including customers and salespeople—don't really understand the role, skills, and functions of salespeople.

- There are a lot of poor salespeople.

To achieve success, salespeople must overcome negative images. They must believe, as I do, that selling is a profession. Properly executed, the skills of professional salespeople are important keys to spreading ideas, educating people, facilitating informed decision-making, providing competition, implementing change, handling commercial details of transactions, and ensuring customer and user satisfaction.

This book is focused on business-to-business selling of material handling equipment and services—forklifts, conveyor, pallet rack, automated systems and much more.

It will:

- Define the functions material handling salespeople perform for customers

■ Describe the environment in which material handling sales-people operate

■ Introduce a sales model for performing effectively in this environment

■ Give specific skills, techniques, and tools to excel

■ Provide a memory tool (a diagram) as a daily reminder

This book respects selling as a profession, and salespeople as professionals.

As far as I know, there has never been a Hollywood depiction of a material handling salesperson.

Let's hope it stays that way!

A Sales Model

Six Areas of Material Handling Sales Skills, Techniques, and Knowledge

To succeed as a material handling salesperson requires developing skills, techniques, and knowledge in six areas:

1. **Product knowledge.** This is the most traditional sales knowledge for material handling equipment salespeople. It involves models, features, application information, specifications, competitive comparisons—details about the "stuff." The danger here is that salespeople sometimes believe product knowledge does the selling. "This stuff sells itself" becomes the cry of some salespeople when new models are introduced with the latest features. No, it doesn't. Product knowledge is only a basic starting point for material handling salespeople. Providing product knowledge training is primarily the responsibility of manufacturers and suppliers of the equipment.

2. **Time management and personal organization.** Managing time and organizing to sell are critical skills. Because many companies incorporate time and organizational tools with database management, customer relationship management (CRM) systems, master calendaring and other systems, salespeople should coordinate their training in this area with their

employer. Time and personal organizational management training is also available from several national firms.

3. **Employer-specific knowledge.** In material handling, the success of the salesperson is tied closely to how well he utilizes the resources of his employer. Time should be spent understanding the employer's strengths, procedures, working relationships, customer service response mechanisms and more. Material handling salespeople need to know who they can go to in their own organization to get things done: to get pricing on specials, obtain information for proposals, process orders, expedite orders, solve customer service issues, get IT help, etc.

4. **Sales technology.** Laptops, voice mail, email, online training, digital order processing, calendaring, database management, customer relationship management (CRM), product configurators, word processing, personal digital assistants, online order status reports and management, business system interfaces, supplier and customer electronic interfaces. Material handling salespeople must quickly become competent and fluent in all these and other relevant technologies as they apply to their specific sales situation.

5. **Industrial geography.** Material handling salespeople must orient themselves to the "who and where" of the customers most likely to be high-potential prospects for their products and services. This includes what job titles are most appropriate for entry-level contact at customers; the physical industrial geography of their assigned area of responsibility; online lists and prospecting tools available; customer and prospect history and database information available at their employer. In other words, salespeople must be able to know where to get answers to the question: "Who and where are all the customers?"

6. **Sales process training and model.** This is training in "what it takes" to create sales in the material handling environment. This is the training and model provided by **Objective Based Selling!**

"He's a born salesman"

This phrase is often used to describe a person who seems skilled at persuading others to do something, believe something, or buy something. It literally says salesmanship is a trait a person is born with, rather than a professional skill which can—and must—be learned.

Other "theories" about sales include:

"Sales is an art, not a science."

"You're either born with sales skills or you aren't."

"Anyone can sell if they just follow the system."

"Sales is a science based on human psychology."

"I sell my own way—I don't rely on any of those sales gimmicks."

"Every sales situation is different. I just go with my instincts."

These statements seem to imply one of two things: either you can't learn selling because it is something you're born with *or* it's all just a bunch of systematized gimmicks—learn those and anybody can do it.

So, is selling a science? Art? Natural skill? Gimmick? System? Individual attribute?

Selling is a profession. Like other professions, there is a body of knowledge that forms the structure of techniques and skills that are most effective. These skills and techniques are practiced in customized ways by individuals with differing levels of effectiveness based on: their knowledge and practice of these skills and techniques; the unique talents they bring to the profession; their circumstances; their work ethic.

Sales model

One method of providing structure to help sales professionals develop their sales skills and techniques in an organized manner is to consistently work with a sales model.

Sales models include:

- An understanding of the characteristics of a specific sales situation (in this case, business-to-business sales of capital equipment—material handling equipment)

- A sales philosophy (examples include: "Nobody beats our price"; "Our products are the best, the Cadillac of the industry"; "We focus on the customer")
- Consistent sales approach
- Sales language
- Memory tool (the diagram!)
- Sales techniques and tools effective in specific sales environments

Objective Based Selling is a sales model to specifically help material handling sales professionals identify, learn, and practice key skills and techniques for their unique sales environment—the business-to-business selling of capital equipment.

This book will present the **Objective Based Selling** model, with recommendations for its effective implementation. It's a framework for material handling salespeople to build on with their unique talents, skills, personality—and work ethic!

About terminology in this book

Men and women excel as material handling salespeople. This book will use the gender-neutral words *salesperson* and *salespeople* instead of the more traditional *salesman* and *salesmen*. The male pronouns *he, his,* and *him* will be used, to avoid the lengthier *he/she, his/her,* and *her/him*.

This book refers often to "business-to-business" selling. This is meant to also include selling to governmental and other organizations which may not technically be businesses, but which act like businesses in the acquisition of material handling equipment.

The word *project* is often used in this book to describe a selling situation, instead of the words *equipment, products, sale,* or *purchase*. This is done intentionally, for two reasons:

- Almost all material handling equipment is purchased as capital equipment. When material handling services are purchased on a major contract basis, the transaction is also often treated as a capital equipment purchase, due to its significance to the customer organization. Businesses acquiring capital equipment often refer to the decision-making and implementation processes themselves as "projects."

■ Use of the word *project* by material handling salespeople elevates the process in the customer's mind and frame of reference. This helps these projects compete for funds within the customer's organization—and generally raises the professionalism of the process. This all works to the advantage of the customer, and to the salesperson using the techniques of **Objective Based Selling.**

Chapter One

Mistakes Salespeople Make

Material handling salespeople often make (and repeat) common mistakes which cost them sales; or when they get the sale, cost them gross margin.

It is sometimes said that the first step in recovery when consistently making mistakes is to acknowledge them. This chapter lists material handling salespeople's most common, costly sales mistakes. The reader is challenged to recognize which of these they most often commit.

Talking too much

Perhaps the most common stereotype of salespeople is "fast talkers." This is because many salespeople talk too much. By dominating the conversation, salespeople appear to believe they can bludgeon the customer, with words, into buying from them. Or, perhaps they feel they can "persuade" the customer with their product features and benefits, and superior logic. Wrong on all counts. In business-to-business sales, customers want to talk, want to share their problems and objectives. Customers want to help the salesperson recommend the right thing by letting them know what's important.

Customers are often frustrated in doing so, because the salesperson is constantly telling them something—often something they don't want or need to hear, or something irrelevant to their situation. Rule of thumb: The customer should talk 80 percent of the time, the salesperson 20 percent. Even better: The customer talks 90 percent of the time!

Not listening

Even when salespeople finally stop talking, they don't always start listening to the customer. Often, they just stop and wait their turn to talk again. Yet, with the right questions from the salesperson, customers will tell salespeople how to sell them. But only if the salespeople are listening—*actively* listening. Listening for: what's important to the customers; who else will be involved in the decision; key objectives and objections; rapport-building opportunities; decision criteria; competition's sales story and more. Taking notes is one technique of active listening; another is restating in their own words what they thought they heard the customer say: "As I understand it, (_____)." Salespeople should examine their practices and determine if they are really listening—or simply *not talking*.

Trying to "prove" their product is better, with features

Material handling customers aren't buying features; they're not even buying equipment. What they are buying are means of reaching their objectives or, in other terms, solutions to problems. Features of equipment only mean something to a customer when they directly address an objective or a need the customer acknowledges. Salespeople will only know which equipment features, if any, are important to customers by asking appropriate questions, and listening to the answers. Telling customers about wonderful features they don't need or know they need, doesn't sell. In fact, if the customer feels he is paying "extra" for features not needed, this relentless listing of features can be counterproductive. "I don't need a Cadillac" is a phrase used by customers who get uncomfortable with all the features being touted by a salesperson.

Handing the customer a brochure—too early and too often

Because material handling equipment is—well, *equipment,* there are lots of brochures filled with pictures, cut-away drawings, and specifications. And salespeople love to hand (or mail or email) a brochure to customers. Early and often. Yet brochures don't sell anything. In fact, the easiest way to get rid of a salesperson is to say, "Give (send) me a brochure and I'll look at it and get back to you." Somehow, it is hard to picture those customers poring over the brochures, then suddenly and excitedly calling to say, "Wow, great brochure. I'll take that!" Most brochures end up in folders or trash cans—usually unread. When customers review brochures, they are often confused by all the technical information, of which only a small part is relevant to their situation. And electronic brochures can be difficult to read, and are even easier to get rid of—that's what delete buttons are for!

Giving the customer pricing information too early in the process

Customers want to talk price. Early in the sales process they ask questions like "What's the price range" or "How much have these gone up?" or "What's a ballpark price?" or "What is the quantity price?" or "What's our discount?" or "What's our national account pricing?" or even "Can you beat this price?" Salespeople often respond with numbers, before they even know enough to give an accurate estimate. *And whatever price given will be the highest price the customer will ever want to pay—no matter what else changes.* In the play "The Price" by Arthur Miller, a used furniture dealer is called in to give a bid to buy furniture in an estate. From the very beginning, he is constantly asked by the family, "What's the price?" He consistently ignores the question—and continually asks more questions about the furniture, the circumstances of the sale, the needs of the family, time frame of disposition. Everything the family reveals works to the dealer's advantage for ultimate pricing. He also is building rapport with the family. At one point the dealer responds to the repeated price question essentially with: *How can I give you the price when I don't know the story?* The same is true for material handling salespeople: Don't give the price until you know the story and have built a relationship. All material handling sales-people should get "The Price" and read it. Before giving another cus-tomer a price!

Focusing the proposal on the price

When salespeople give customers proposals (all too often called "quotes"), the customer almost immediately goes to "the price." And when the customer sees "the price," selling stops. The customer is mentally comparing the price numbers to budgets, previous pur-chases, decision criteria, and competitive prices. Everything but focusing on why the salesperson's proposal will help them meet their objectives. Salespeople must learn to write proposals that de-emphasize price, presenting it finally in a format and perspective that is favorable to the salesperson's story. Salespeople must also learn to delay price considerations and questions during personal proposal reviews, until they have "set the stage" for price consider-ation and presented their proposal effectively.

Offering and conducting poorly timed, unfocused, ill-advised demonstrations and site visits

Because material handling salespeople sell equipment that can be seen and operated, or controls, software, or methods which con-trol equipment or tangible things, natural requests of customers include:

"Can we see one?"

"Can we try it out?"

"Can we visit an operation where it's in use?"

And, a natural response for the salesperson is "Sure!"

Especially with forklifts, salespeople often beat customers to the punch, offering a demonstration before the customer even asks: "Let me bring one out and let your operators try it out." This seems to be an attempt to emulate the car sales mantra of "Get 'em in the dri-ver's seat and you can close the deal."

With larger, installed material handling equipment which is not as transportable (conveyor, vertical lifts, mezzanines, storage systems, etc.), the salesperson's offer often is: "Let me take you to show you XYZ Corporation, which has one, or which is operating that equipment." If this is not easily done, salespeople offer to show a video or give a case history.

With software or complex systems, a visual and data simulation is sometimes offered. Or a visit to a demonstration center.

The problems with many of these demonstrations, site visits, or simulations are:

- They can be time-consuming and expensive to arrange.

- Because salespeople and customers are not clear on the objectives of the demonstration/site visit/simulation, no one knows clearly if it was "a success." What did it really show that will help close the sale or convince the customer that this equipment meets his objectives?

- Because demonstrations can be difficult to arrange and schedule, and may not take place at the customer's location, it is common that not all the customer decision influencers are present. So it either must be repeated, or it simply doesn't persuade a key person who is not able to be there.

- Decisions to purchase material handling equipment have an extended time frame. The subject of demonstrations or site visits often comes up early in the sales process, leading to a significant time separation between demonstration and decision time. When the demonstration takes place early in the sales process, the full parameters of the customer's objectives and situations are often not known. As a result, the wrong equipment may be demonstrated, an irrelevant site visited, or the wrong points emphasized during the process.

- Equipment often has a problem and does not perform as expected during demonstrations or site visits. Oops.

The list of factors causing ineffective demonstrations and site visits is long. Demonstrations and site visits can be effective tools for material handling salespeople, but only when they are strictly controlled, with a focus on clear customer objectives and parameters; with the right people present; and are conducted at the appropriate, pivotal time in the sales process. The **Objective Based Selling** model offers specific guidelines on how to conduct effective, *Objective Based*, customer-focused demonstrations/site visits/simulations.

Summary

We've now seen some of the most common, costly, repeat mistakes that material handling salespeople make. **Objective Based Selling** will give salespeople the tools and techniques to get customers talking; to actively listen to their objectives and concerns; to stop boring customers with brochures they don't read; to avoid talking price early in the process or the proposal review; to prepare proposals that present price in the context of a total sales proposal; and to avoid conducting unneeded or ineffective product demonstrations.

Chapter Two

The Real Job Functions of Material Handling Salespeople

Salespeople understand their function is to produce sales for their company. While true, it is not a function that customers value. This chapter examines the customer-focused functions of material handling salespeople. The **Objective Based Selling** model gives salespeople the tools and techniques to accomplish these functions—making them more highly valued by customers and thereby leading to more sales at higher gross margins.

Help customers understand and define their own situation

Material handling salespeople bring fresh eyes, outside perspective, and material handling expertise to their customers' material handling operations. They see things customers may overlook or misunderstand. Things like unsafe practices or equipment; operator abuse resulting in excessive maintenance costs; wasted space; outdated inventory management practices; use of old technology, adding costs to customers' operations; energy-inefficient situations; poor material flow; equipment being used beyond capacity, etc. In material handling and facilities operations, there is constant room for more effective, safer, more space-efficient operations. By asking the right questions, physically touring operations, reviewing appropriate customer management reports, and talking to operators, supervisors, and others in customers' operations, material handling salespeople can help customers understand the areas where improvements can be made. It's like holding an educated mirror up to the customer's operations so they can see what's really happening. This encourages the customer to look to the salesperson—who reveals these situations—to provide products, services, and solutions for improvements.

Introduce appropriate new ideas

A traditionally acknowledged function of salespeople is to introduce customers to new ideas: new equipment, new technologies, new methods, new insights, new trends. However, to avoid wasting customers' time on inappropriate topics, material handling salespeople must first learn about their customer's operations before "throwing out ideas." The key is to introduce *appropriate* new ideas to customers, ideas that will help customers achieve their objectives.

Help customers make choices from among alternatives and options

There is a school of thought that consumers—corporate decision-makers—are suffering from "decision overload." There are so many ways of doing things—alternatives, models, options—that choices that were previously considered simple decisions are now multiple decisions to be made among an increasing array of confusing alternatives. As in many fields, all aspects of material handling have exploded with new technology, creating multiple alternatives. A key function of the material handling salesperson is to help customers choose the most effective products, services, and options for their operations, from an increasingly sophisticated group of choices. This can only be done skillfully if the salesperson first understands key elements of the customer's objectives and operations, so that he can make appropriate recommendations. Ideas must be communicated in a manner the customer understands and agrees is right for their operations. This leads the material handling salesperson to focus first on customer operations before beginning detailed discussions of equipment and alternatives.

Facilitate decision-making

Most significant material handling sales are made to companies or organizations: business-to-business selling. Several factors make it difficult for organizations to make timely, appropriate decisions. These include:

- Multiple decision influencers: each with a different perspective, knowledge of the situation, responsibility for the outcome, time frame, priorities and so forth.

- The proliferation of alternatives, competitors, options, and methods can make it difficult for companies to decide on what's right for them—and then act on it.

- Larger organizations increase the levels of approval required to reach a decision, or get a decision approved.

- Nationwide and international customer operations increase the levels of review, and often introduce decision influencers who are not familiar with local situations or urgencies.

- Corporate purchasing agreements may limit choices, excluding the most appropriate local choices for national agreement reasons.

- Complicated corporate budgeting and capital equipment justification processes. These may not be totally understood by the users or those managing a specific project, making it difficult for them to "sell" the project upward.

- Competition for funds in the customer's organization.

- Ineffective selling by the primary customer contact for a project, within his own company.

■ Significant dollars of many material handling projects can slow the decision process, simply by additional review required or reluctance to commit these dollars or take the risks involved in implementing the project.

From the initial contact with a prospect, material handling salespeople should be thinking about how they can help the customer facilitate their decision-making. **Objective Based Selling** will provide questions to uncover the customer's decision-making process and influencers for given projects; proposal formats to help customers identify how a given project will ultimately benefit them so they are comfortable moving forward, and which formats help sell to remote decision influencers the salesperson can't meet; skills to obtain and effectively conduct "scrum" meetings of multiple customer decision influencers, to move the process forward; and strategies for effective, customer-focused demonstrations/site visits/simulations. The salesperson who is most skilled in helping customers reach decisions is often the salesperson who most often gets the deal.

Help implement decisions

Successful operation of new material handling equipment, or implementation of larger projects, usually requires adjustments as the delivery or project proceeds. Material handling salespeople are expected by both customers and their own company to help implement decisions to buy their products or services. This begins at the time of sale with the salesperson accurately communicating the details of the customer's project, selection, and circumstances to their own company and suppliers (in other words, accurate, prompt order processing, work order preparation, and communication). It includes anticipating issues and dealing with changes as well as with the unexpected. Helping with decision implementation may be as simple as getting something delivered today or as complex as working closely with project managers on integrated sales over many months. It may include delaying or expediting delivery; physical modification of the product; reprogramming software; training operators; even changing the financial transaction details—perhaps from purchase to lease at the last minute. It is the responsibility of the material handling salesperson to help implement the decision to the customer's satisfaction. The good news about this is that because material handling sales are often "repeat business," successful implementation of a project can be the start of a new project or the source of a referral.

Follow through

Customer satisfaction with the completed project is the function and responsibility of the material handling salesperson. Following through in the material handling industry means more than just "doing what we said we would do." The current, higher standard is assuring the project "does what the customer thought he heard us say it would do." Measures of the success of follow-through are: on-time payment in full by the customer; the eagerness of the customer to do business with the salesperson again; willingness of the customer to give the salesperson referrals to others who might need his services.

Handle commercial details

As the primary interface between his company and the customer, the material handling salesperson handles commercial details as one of his key functions. These details include credit approval; accurate order transmission to their company's business system; helping their company meet customer billing requirements; handling paperwork (such as lease documents) for finance transactions; obtaining the required purchasing paperwork from the customer; handling requirements for down payments or progress payments; where required, helping their own company and their customer deal with contracts; understanding customer payment systems (check, electronic funds transfer, credit card); dealing with customer requests for returns—and helping manage the return if approved or informing the customer if not approved; and, in many companies, helping with collection of past-due bills. In handling commercial details, the material handling salesperson is "two-faced"! The salesperson represents his company's interests to the customer, and many are issues where the salesperson represents the customer's interests and requests to his company.

Paid to perform functions for the customer

"Estimators" or "bidders" can respond to customer requests to bid or supply products, often without leaving their computers! Material handling sales professionals, on the other hand, recognize they are paid to provide the functions listed in this chapter for their customers. It's the salesperson's "value added." As their relationship with customer contacts increases and rapport is built, customers also recognize this, and gain trust in the salesperson's ability. That's when customers react favorably to a salesperson's recommendations and proposals. That's when salespeople sell more at higher gross margins!

Chapter Three

Know Your Sales Environment

Salespeople sell most effectively when they understand and can articulate underlying characteristics of their sales environment. They recognize "what's really happening" in specific sales situations, and take actions most likely to lead to a profitable sale.

For a salesperson to consistently use a sales model, he must have confidence that the model provides tools to effectively deal with the sales situations he faces on a daily basis, as well as over the entire sales cycle.

This chapter describes fourteen characteristics of the material handling sales environment, along with a brief discussion of implications for the salesperson's actions. These characteristics will be referred to again and again as the **Objective Based Selling** model is explained. Specific sales techniques are recommended to deal with this environment.

1. Multiple decision influencers

Material handling sales of consequence almost always involve more than one decision influencer. In large corporations, depending on the size of the project, these may include local operations managers such as warehouse or manufacturing managers, and persons

with responsibilities such as safety, budgeting, finance, maintenance, purchasing, contracting, general management... it can be a
long list. In some corporations, even equipment operators are given
the opportunity for input, veto, or primary selection influence.
Generally, the larger the project or the more complex the organization the salesperson is dealing with, the more people involved in the
decision. Knowing this "going in" alerts the salesperson to pay attention to everyone he meets; to look for opportunities to meet more people; to ask questions to identify decision influencers and the roles
they play. In smaller companies, salespeople may be dealing with
owners who seem to have discretionary decision-making power.
However, even owners usually get input from others—perhaps a
family member or banker not directly involved in the business.

2. Hidden decision influencers

No matter how skilled, salespeople often fail to identify all the
decision influencers. And in many situations, salespeople will never
get to personally meet some decision influencers, whether identified
or not. Material handling salespeople must assume there are "hidden" decision influencers in every project of consequential size or
complexity. This creates the salesperson's challenge of "selling"
influencers who cannot be identified or met. There are two effective
ways of doing this: via the written proposal document, which com-
municates to those decision influencers salespeople can't meet; and
through the salesperson's "advocate" (sometimes call his "angel").
The "advocate" is the salesperson's primary contact with whom a
personal, professional relationship has been developed, and who
believes his company should act on the salesperson's proposal.

3. Access to decision influencers increasingly difficult

Customer organizations erect barriers to salespeople's direct
contact with decision influencers. Barriers include phone systems
(voice mail!); gatekeepers; national contracts; physical distance;
email filters; policies; and simple refusal to see salespeople. This
places a premium on communication and relationship-building
skills when a salesperson is able to make personal contact. It also

increases the importance of effective written (email and paper) communication—particularly the proposal document. While the salesperson must work for the opportunity to personally review his proposal with customer decision influencers, he can't count on it.

4. Salesperson often not in the room when purchasing decision is made

Traditional sales training often emphasizes "on the spot" closing techniques. This assumes, of course, that the salesperson is dealing with a final decision-maker, and that the decision-maker is authorized and willing to make a decision on the spot. This is most often **not** the case with material handling projects of consequence. Things working against "on the spot" decisions (and closes) include multiple and hidden decision influencers; complex corporate structures and decision criteria; the size of many transactions; the impact these decisions may have on customer operations; simple reluctance of many people to act "under pressure" from a salesperson. When salespeople are precluded from closing in person, they must develop written proposals which sell when they can't be there; relationships with customer advocates who will make their case; and proposal support techniques to justify post-proposal contact.

5. Competition for funds

Material handling equipment most often is purchased with capital equipment budget funds. Inevitably, there are more requests for capital funds than funds budgeted or available (hence the jargon term for capital equipment budgets: "wish lists"). In many material handling projects, the case must be made that doing your project is more valuable to the customer than other identified needs. To help his project effectively compete for funds, the salesperson must understand the customer's objectives for the project; its financial return to the customer; and why it's important for the customer's operations that it be done. This must be communicated to multiple and hidden decision influencers through the written proposal and the salesperson's advocate. This is also important in dealing with the next characteristic...

6. A major competitor is often: Do nothing!

One alternative for many material handling projects is to do nothing. If projects don't meet internal financial criteria, or don't have a persuasive advocate, they are often simply dropped. To the material handling salesperson, this has the same negative effect as losing an order to a competitor. The salesperson has invested time and money in preparing a responsive proposal and gets nothing but a "Thanks for your help" in return. To avoid this, salespeople must be strong at qualifying projects, so they avoid investing time in projects that don't make sense and aren't really going to happen. It also means they must develop skills to learn a customer's internal financial and other criteria for implementing projects.

7. Sales time cycles not instant

Time cycles for evaluating, purchasing, and implementing material handling projects are not "instant." Typically, from initial sales contact to purchase, time frames may be from one month to over a year. Some complex projects may take up to three years or longer to develop, corresponding with budgeting cycles, construction schedules, or other corporate time frames. Salespeople must develop skills and techniques to deal not only with the daily demands of personal sales contact, but also to advance projects over longer time cycles. Understanding a customer's "pace" in a project is key. Another important skill is identifying and dealing with change over extended time frames. As weeks and months go by, there may be changes in customer decision influencers, priorities, application parameters, and budgets. There may also be change in things such as the models, brands, specifications, and lead times of products being offered by the salesperson. Personal relationships and questioning skills are critical to dealing with change over extended time sales cycles typical of material handling sales.

8. Significant dollars often involved

Most material handling projects involve dollars which are significant to customers, and involve more complex purchasing processes and higher levels of scrutiny than smaller purchases of lesser impact. Higher dollars means more and higher levels of decision influencers; longer time cycles; more competition for funds. This is often also related to the next characteristic...

9. Career decisions being made

Decisions made by customer contacts in material handling projects are often career decisions. These are decisions involving significant dollars. Because changes to company operations get high visibility, the project results may advance or retard the customer champion's career. If projects really don't work out, or aren't perceived to meet objectives, careers may abruptly end! This can be particularly highlighted when an advocate recommends changing brands or methods. When career decisions are being made, one key element usually trumps others: trust. If the salesperson cannot build trust that the project will succeed and reflect positively on the champion, he is unlikely to get the business. If the salesperson is dealing with an owner, something even more important than just a career may be at stake: the customer's company profitability or viability. While the customer may constantly say "price" is a primary criteria, the material handling salesperson knows—within some pricing range—"trust" is the real criteria.

10. National purchasing increasing

As more corporate mergers happen and more purchasing control is centralized nationally (and internationally), many of the characteristics listed above are exaggerated: more decision influencers—many of them hidden—with more scrutiny and competition for funds. Not only does this increase the importance of written proposal document, but skills such as the ability to find alliance partners in either the purchasing or using location come into play.

11. Brands decreasing in importance

There was a time in material handling sales, particularly sales of lift trucks, loading dock equipment, conveyor and other branded equipment, in which brands drove decisions. If you sold the right brand, that was 80 percent of the battle. No more: Decision criteria are now more often driven by internal financial considerations; real, hard objectives to be accomplished; and multiple, hidden decision influencers who have no brand allegiance. Sales skills in dealing with these situations are now more valued than "What brand are you selling?"

12. Alternative solutions available

There are almost always multiple, alternative methods and equipment to accomplish customer objectives. Alternatives may include used versus new equipment; cash versus financed purchase; a myriad of finance alternatives; conveyor versus lift trucks versus automation versus manual; and so on. Where there are multiple objectives to be accomplished (often the case), the chosen solution may depend on how the objectives are negotiated and prioritized within the customer's organization. In other cases, the choice may depend on which parameters are considered, or the competition for funds. To deal with this, material handling salespeople must be skilled in helping customers negotiate priorities; uncover parameters; make decisions. Salespeople should be ready to offer alternative solutions, at different price points, so that customers are choosing between two of *their* solutions—not theirs and a competitor's. Salespeople must be able to articulate how the benefits of alternative solutions impact customer objectives. Salespeople must be able to effectively conduct scrum meetings with multiple customer decision influencers.

13. Semi-technical field

Material handling is a "boutique" industry with its own jargon and assumptions. Compared with Information Technology (IT), material handling is semi-technical. As IT solutions are incorporated more into material handling equipment and methods, it is becoming more technical. Salespeople must be careful not to use "insider" or "technical" jargon which loses the customer. Without talking down to customers, salespeople must be sure all terms used are fully understood by the customer decision influencers. This is particularly important when dealing with equipment specifications, or customer parameters, on which orders will be placed.

14. Customer decisions are often political

A common background for material handling salespeople is engineering. Engineers tend to believe there is one, "provable" solution to a problem or project. Any material handling salesperson looking to "prove" that their proposal is the best—through the use of figures, references, proof statements and the like—is doomed to ineffectiveness and failure. While these all play a role, significant decisions involving more than one person (hence, almost all decisions to purchase material handling equipment) involve the personal and power relationships between the people involved. Salespeople who understand this and develop the questioning, personal relationship and communication skills to deal with it will have long, productive careers!

The fourteen characteristics discussed in this chapter help define the sales environment within which the material handling salesperson operates. **Objective Based Selling** was designed specifically to provide the material handling salesperson the tools and techniques to be effective in this environment.

Chapter Four

Learn the Objective Based Selling Language

Selling involves verbal communication and mental activities. The right language can help create a positive selling environment—for both the salesperson and for the customer. Conversely, the "wrong" language can provide a negative selling environment. An example often used is the difference between using the words *price* (or worse, *cost)* and using the word *investment*.

Since *price* and *cost* indicate things which are basically negative, their use should be minimized. *Investment* indicates the positive improvement of a business through purchasing (investing in) equipment or services that will offer a return or "pay for themselves." This is not just a play on words; in fact, the purchase of material handling equipment or services had better be helping the customer's operations, paying for themselves in improved productivity, better space utilization, lower maintenance costs, increased output, more accurate order fulfillment, energy savings—or the equipment shouldn't be purchased! And, use of the word *investment* sets a positive tone.

Below are eleven additional word choices recommended to help create a positive sales environment, along with a brief explanation of why the recommended words create a more positive environment than the commonly used alternatives shown in parentheses.

1. *Objectives* (versus *problems*)

Many sales models talk of "problem solving selling" and encourage salespeople to help customers identify and "solve" problems. However, a more positive activity in many situations is to help customers meet objectives rather than simply overcome problems. Customers, while often reluctant to admit problems, are almost always ready to discuss objectives to be achieved. The very question "What problems do you have?" indicates the customer—or somebody—is doing something wrong (Who should we blame?). A better question: "What are your primary objectives for improvement?"

2. *Decision influencers* (versus *decision-makers*)

Advice for salespeople used to go like this: "Find the decision-maker and focus your attention on him." However, as indicated earlier, the material handling sales environment almost always involves multiple decision influencers, not just one decision-maker. Salespeople who ask customers to identify a decision-maker are subtly demeaning others in the process and deluding themselves into thinking if they can just find that one guy, they can close the deal with him. However, by asking, "Who are the decision influencers in this project?" the salesperson has the likelihood of identifying and meeting more of them and less likelihood of overlooking or slighting a key decision influencer. Even when salespeople are dealing with individual owners of companies, who might appear to be "ultimate decision-makers," these owners often rely on others for advice—or defer to them for decisions. These are the decision influencers the salesperson must identify, and if possible meet, learn from, and influence. A further benefit of the term *decision influencers* is that customers are more comfortable identifying them rather than a decision-maker. It's less threatening to identify decision influencers.

3. *Parameters* (versus *specifications*)

Material handling salespeople love to talk about the specifications of equipment they are selling; of competitors' equipment; of customers' current equipment, and customer specification requirements. This can lead the salesperson quickly to recommend specifications that are "common" or "standard" or "the same as you have" or "superior to the other guy," rather than identifying what really meets the customer's current situation and objectives. The better focus is to identify critical parameters of the customer's job. Then, and only then, can appropriate equipment specifications be recommended. Examples: For forklift masts, determine the highest lift requirement, lowest facility and loading restrictions (operational parameters), rather than listing mast specifications. For conveyor, determine the most common and extreme box size and weight measurements (parameters) before specifying frame widths, roller spacing, belt type, etc. And so on. Additional bonus: Customers constantly hear salespeople discuss "specifications"; they'll take notice when a salesperson uses the much less common "parameters"!

4. *Time frame* (versus *delivery* or *"When do you need it?"*)

A discussion of time frame with customers (such as "What is your time frame for reaching a decision on this project?" "For implementation of this project?" "For getting approval?") is a more professional discussion than the commonly used "When do you need delivery?" Additionally, there may be more important time issues for the customer than delivery, and this focus on time frame encourages the customer to tell the salesperson the most important thing to them about time.

5. *Financial considerations* (versus *budget*)

In material handling projects large and small, the word *budget* usually comes up. Of course, a budget by definition is a constraint, a restriction on what can be spent. It also implies the budget number is the only or the most important financial issue in a project. Instead of asking a customer about their budget, or responding directly to their use of that word, it is more effective for material handling salespeople to ask, "What are the financial considerations in this project?" This encourages the customer to talk more broadly about money, rather than just its constraint. It also more easily allows the introduction of the concept of financing and payments as a financial alternative to a capital budget constraint.

6. *Proposal* (versus *bid* or *quote*)

A bid or a quote is a number. Low bid wins. That's not the path to higher gross margins and bigger commissions. Additionally, it is not to the benefit of the material handling salesperson to give the impression his products or services are commodities, which is the implication of the terms bid or quote. They imply all offerings are the same, and price (the bid) is the only important thing. Material handling project implementation, on the other hand, involves warranties, life span, rate of return, trust that the job will be completed to customer satisfaction, appropriate selection from alternatives, follow-through—it's a whole way of doing business. Material handling salespeople should present customers with proposals that outline how their offering and way of doing business helps the customer meet their objectives, within the customer's parameters, in a distinctive manner, with implementation and follow-through assistance by the salesperson and his company. Salespeople should not give customers a quote. Quotes are an expense; proposals describe investments.

7. *Review* versus *presentation* (often heard as *sales pitch!*)

Most customers don't want to hear a salesperson give a presentation—even a presentation of a proposal. They assume it will be a mind-numbing, self-promoting sales pitch. And they're probably right! Rather than asking for time to "make a presentation" or "present the proposal," better to ask to get decision influencers together to "review the proposal, confirm objectives and parameters, and answer questions."

8. *Scrum meeting*

A distinctive term which can be used for a meeting where decision influencers are assembled to meet with the salesperson is a "scrum meeting." Customers can often be persuaded to pull together the decision influencers for a project, with the salesperson, in order to:

■ Be sure everyone's on the same page

■ Identify key objectives and parameters

■ Get answers to questions about recommendations and proposals

■ Prioritize objectives

■ Explore alternatives

■ Facilitate the decision

In a rugby scrum, there is initial confusion and infighting. Then, a leader emerges with the ball and everyone follows that direction. That's a mental image and analogy for the type of scrum meeting skilled material handling salespeople should aspire to with customers. Its stated purpose should be to get everyone involved in the decision headed in the same direction. There are more details about the why, when, how, and implementation of scrum meetings later in this book. For now, practice saying, "What can we do to schedule a scrum meeting to review the proposal with key decision influencers?" instead of "I'd like a meeting to present our proposal."

9. *Project* (versus *equipment purchase*)

As mentioned in the book's introduction, the word *project* elevates a purchase from a simple price-based transaction, to a situation requiring more study, professional expertise, and attention. Use of the word *project* can help the primary customer contact compete for funds with other projects—or purchases—in their company. A further effective tool is to make it a customer-focused project name, instead of a product-focused name. Thus, a mezzanine purchase becomes a "Warehouse Floor Space Expansion Project" or "Space Optimization Project." A purchase of forklifts can become a "Forklift Fleet Modernization Project" or a "Forklift Expense Reduction Project." A loading dock trailer restraint acquisition should be a "Loading Dock Accident Prevention Project" or "Dock

Safety Upgrade Project." This terminology can be used by the sales-person in conversation and correspondence or email communication. Many customers will notice and adopt the terminology. This terminology should be used on all proposals.

10. The inclusive *"we"* (versus the adversarial *"you"*)

The preferred position for a material handling salesperson is to be on the same team with the customer, helping achieve objectives. As early in the relationship as may be appropriate or comfortable, salespeople should begin using the inclusive *we* and *our,* as in:

- "What are *our* objectives in this area?"
- "When do *we* need to be operational?"
- "Where are *we* short of space?"
- "What's *our* time frame"
- "What are *our* financial considerations?"

These phrases may not be appropriate in an initial meeting with a customer. They might imply a relationship that doesn't yet exist, and be perceived as presumptuous. But as relationships with customers develop, the use of *we* and *our* is acceptable language to create a positive sales environment and break down adversarial attitudes.

11. *Gross margin* (versus *mark-up*)

These are terms regarding pricing, which are not usually used with the customer in the room. They are terms used by salespeople and companies as pricing strategies are managed. While often used interchangeably, they are not the same. The math is different. *Mark-up* is a commonly used term in some industries and is simpler for most salespeople to figure. They simply take their cost and multiply it by **1. "mark-up."** Example: A 20 percent mark-up on $1,000 of cost is calculated by taking $1,000 x 1.20 to equal a selling price of $1,200. However, most material handling companies operate with targets for gross margin. Gross margin is calculated by dividing the cost by the inverse of the target gross margin. For example, to get a selling price with 20 percent gross margin on an item with a cost of $1,000, **divide $1,000 by .8** to equal a selling price of $1,250. That's

25 percent more profit on a sale that might be referred to in either case as a 20 percent sale. What's the point? First, gross margin is more commonly used in financial analysis in the material handling business, so using that term and the associated calculation is more consistent within the industry and easier to compare. Perhaps just as important, earning a higher price with a customer is often an exercise of mental discipline by the salesperson. And, salespeople tend to think certain profit percentages are fair or achievable. If a salesperson is fighting for 20 percent, it's better to be fighting for 20 percent gross margin than 20 percent mark-up!

Chapter Five

Understand the Goal of Objective Based Selling

The goal of **Objective Based Selling** is to provide a model for material handling salespeople to sell more at higher gross margins.

Simply selling more has a fairly straightforward formula: Make more calls and have a lower price.

Selling is indeed a numbers game to a certain extent. Everything else being equal, the salespeople who work harder and who consistently make more contacts will almost always sell more. However, there is a practical limit to the number of potential sales contact hours in the day. There is also a limit on the time a salesperson can devote to customer contact (making calls), due to the follow-through, training, paperwork, and other requirements of the job.

Salespeople must learn how to be more effective on the calls they do make, rather than just make more. It's the old adage, Work smarter, not harder. Or, as a sales manager might say, "Work harder—and smarter!"

As to selling at a lower price:

- Why do companies need salespeople to drop price?

■ Someone always has a lower price—by selling something different, redesigning a product or project, cutting margin …

■ In material handling sales, the salesperson's income, and his company's financial health, is almost always tied to gross margin. Even a slight increase in gross margin can leverage into significantly more commission, higher job performance ratings, and better company financial performance.

■ Where's the fun and challenge of selling a lower price?

Objective Based Selling is designed to help sell more at higher gross margins by changing material handling salespeople's traditional focus on product, to the more effective focus on customer.

Objective Based Selling is designed to change the selling process focus from the product to the customer.

Because material handling products and services involve tangible things—equipment—material handling salespeople (and some customers) naturally focus on the stuff: its specifications; its appearance; its operation; its brand name; its features; its delivery; and, of course, its price.

Yet customers only purchase material handling products and services as a means to an end, a way of accomplishing objectives in their operations. And in the end, most are a lot less interested in the stuff than the salesperson thinks.

Customers make business purchasing decisions for two sets of reasons: to accomplish the business objectives of their immediate situation; and to satisfy their personal objectives. The consistently successful salesperson needs to understand these objectives and show the customer how they can accomplish their objectives by acting on the salesperson's proposal.

The "objectives" in **Objective Based Selling** are the customer's objectives. The focus should be on the *customer's* objectives. The customer is the most important part of the sales process. The customers are the ones spending the money!

If salespeople meet the right decision influencers, ask the right questions, listen carefully to answers and ask follow-up questions, observe, and build rapport with customers, the customers will tell the salesperson how to sell them.

Chapter Six

The Four Keys of Objective Based Selling

The essentials of this sales model are embodied in the **four keys** of **Objective Based Selling,** which are:

- **Open-ended questions**
- **Personal, professional relationships**
- **Customer-focused proposals**
- **the Objective Based Selling diagram**

Key number one: Open-ended questions

Objective Based Selling provides over 100 specific open-ended questions to help the salesperson:

- prospect, and qualify prospects
- encourage the customer to talk about their company's objectives, operational situation, purchasing procedures and criteria, decision influencers, and personal objectives

- understand the customer's situation in order to make responsive recommendations
- create forward motion in the sales process
- build trust with the customer
- perform critical sales functions for the customer
- build rapport with customer decision influencers

Key number two: Personal, professional relationships

While many forces in today's material handling sales and purchasing environment are pushing toward depersonalization, in the end all purchasing decisions are made by people. People still buy from people, even if they use computers and national contracts to do it.

The larger the project, the more critical the purchase is to the customer, the more important it is for the material handling salesperson to establish a personal, professional relationship with key decision influencers. These relationships will provide:

- access to other decision influencers
- information leading to responsive recommendations which are customized to the customer's objectives and parameters
- opportunities to build trust
- better understanding of the customer's decision-making criteria and process
- coaching for the salesperson in how to deal with the customer organization
- advocacy for the salesperson's proposal
- the opportunity to provide a modified, more targeted proposal at the time the customer is really ready to act. Sometimes called the "second chance," this will be discussed in more depth in chapter twenty, "Modify Proposal," an element of the **Objective Based Selling** diagram
- early warning (and opportunity) of the need to react, adjust, recover during the implementation phase of a project, when things are not going exactly as planned

Personal, professional relationships are so important that a corollary to the **Objective Based Selling** concept mentioned earlier in this chapter is: **When business and personal objectives conflict, personal always wins.** No matter how responsive the salesperson's proposal is to customer objectives; no matter how attractive the price; no matter how strong the salesperson's proposal; if, in the end, the customer does not like, believe, understand, and trust the salesperson and his company, the customer will find a way to purchase elsewhere.

Key number three: Customer-focused proposals

All material handling purchases (sales) of consequence involve a written proposal (paper or electronic or both) before the customer makes a commitment.

Most proposals by material handling salespeople have two critical flaws:

- Material handling proposals usually focus primarily on the equipment, service, methods, or software being proposed. In the extreme, many material handling proposals are simply a listing of equipment specifications with a price quote and delivery statement. No selling, just quoting.

- Material handling proposals often contain no reference to customer objectives and parameters. Without this information there is no indication the salesperson understands the customer's objectives and parameters, or took them into consideration when creating the proposal. Without this information, there is no indication that acting on this proposal will enable the customer to achieve their objectives. Without this information, it's just another equipment quote. Lowest quote gets the order. Or, the project dies due to not competing effectively for funds.

In other words, most material handling equipment and service proposals quote rather than sell.

This book will present the concept of effective, customer-focused sales proposals. It will examine the audience for sales proposals (those multiple and hidden decision influencers), and give specifics about the essential elements of customer-focused, objective-based

proposals. Templates for customer-focused, objective-based proposals will be provided. Templates that sell in the meetings which the
salesperson cannot attend, speaking to decision influencers the
salesperson never meets.

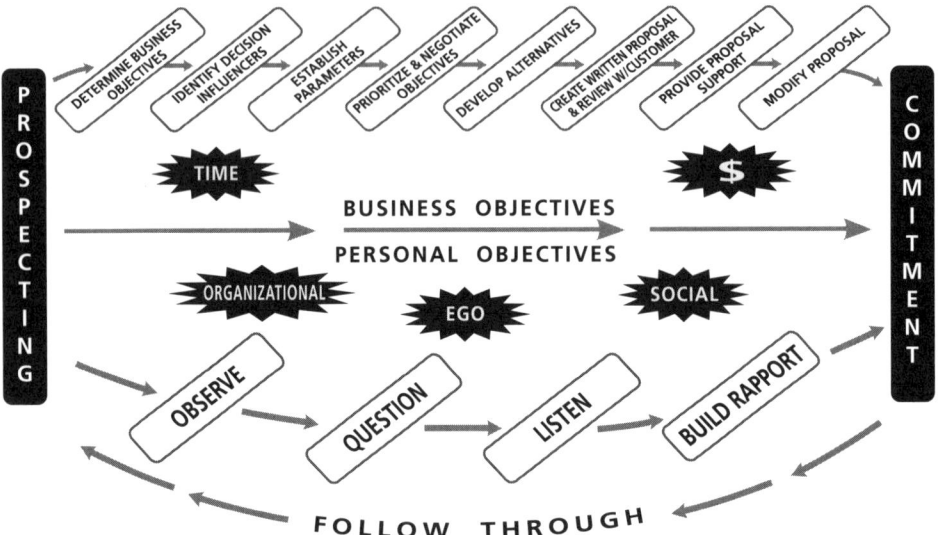

Key number four: the Objective Based Selling diagram

This diagram is a visual representation of the **Objective Based
Selling** model.

The basic idea of **Objective Based Selling**, reflected in the diagram, is that:

> *Customers make material handling purchase decisions*
> *for business and personal reasons—to accomplish busi-*
> *ness and personal objectives. The job of the material*
> *handling salesperson is to determine the customer's*
> *business and personal objectives, and convince the cus-*
> *tomer they can accomplish these objectives by acting on*
> *the salesperson's written, customer-focused proposal.*

Small, medium, and large projects

Material handling sales involves purchases and projects of varying sizes. On the small end of the continuum, the purchase may be for a simple pallet jack, set of wheel chocks, conveyor stands, or hand chain hoist.

At the high end, a material handling project involving a fleet of forklifts, yearly maintenance contract, automation with palletizers and unit load cranes, may involve hundreds of thousands—even millions of dollars.

There are projects of size, complexity, and dollar levels everywhere between these extremes.

The principles and four keys of **Objective Based Selling** are effective at all levels of projects and purchases for material handling equipment and services.

The difference is in the emphasis, and time line, of the different elements and stages of the process.

If the project or purchase is small, the customer will normally not want to devote as much time to getting it done. The salesperson must therefore be ready with questions to move quickly through the process. In a few cases, written proposals will not be asked for, or actually may be asked for after the purchase, to confirm details.

Scrum meetings are not as likely with smaller projects—or may take place spontaneously, standing up in a customer facility. And, of course, there may be fewer decision influencers involved, and less time to build rapport.

Salespeople should always be looking for the opportunities of a small project being the "foot in the door" for future, larger projects. They should not be taken lightly. A well done, customer-focused one-page proposal for a small project may be noticed by the customer and trigger a call back for larger projects. Doing a professional job on a small project is a way to build trust. And, what is small for one salesperson may be large for another.

So, in smaller projects, the emphasis is on the questions and on using the opportunity to build personal, professional relationships.

In larger projects, the elements of the **Objective Based Selling** process are usually more distinct, taking place over a larger period of time. There are likely more decision influencers, opportunities (and necessities) for scrum meetings, reviews of the proposal by hidden decision influencers, and changes over time.

In larger projects, there is more emphasis on the proposal and the personal, professional relationships which allow for opportunities for proposal support and modifying the proposal. Of course, an effective proposal cannot be prepared without those open-ended questions early in the process.

In summary, the four keys of **Objective Based Selling** are:

- **Open-ended questions**
- **Personal, professional relationships**
- **Customer-focused proposals**
- **the Objective Based Selling diagram**

These four keys to **Objective Based Selling** will now be examined in detail.

Chapter Seven

Focus on the Questions

Open-ended questions are the most powerful tool available to the material handling salesperson. Effective use of open-ended questions:

- Focuses the conversation on the customer.
- Gets customers talking about their operations, objectives, parameters.
- Encourages customers to talk about themselves.
- Helps the salesperson stop one of the biggest salesperson's mistakes: talking too much.
- Controls conversations. While it may be counterintuitive, the person asking the open-ended questions is usually the one controlling the conversation. Customers (and people generally) like to answer questions (particularly questions about themselves and their areas of interest). This puts customers in the position of talking about what the salesperson asked about.
- Shows genuine interest in the customer and their operations and objectives, helping to build trust and rapport.
- Helps the salesperson learn about the customer's operations,

so they can perform their functions of helping customers meet objectives, choose from among alternatives, offer appropriate new ideas and so on.

In the **Objective Based Selling** model, questions propel the sales process toward a decision and commitment to the salesperson's proposal. They are so important that throughout the remainder of this book, the open-ended questions recommended for use by material handling salespeople are in quotation marks and boldfaced. They are also summarized in the appendix, sorted and organized by stages of the **Objective Based Selling** diagram, a memory tool for the questions.

This chapter will cover:

- Basics of open-ended questions and questioning
- Funnel theory of opened questions
- Following the trail of the answers
- Customers' reactions to questions
- What to do after asking open-ended questions

Open-ended questions: The basics

The most basic definition of an open-ended question: A question that cannot be answered yes or no.

Questions which can be answered yes or no are called closed-ended questions. And often, they "close" the discussion with the customer, without revealing information that is helpful. A typical closed-ended question for customers:

"Do you have any material handling needs?"

Typical customer answer:

"No."

Now what?

Open-ended alternative question:

"If you could change one thing about your material handling operations, what would it be?"

Open-ended questions get the customer thinking—and discourage quick, trite responses.

So, how do we recognize open-ended questions? The British author Rudyard Kipling helped us with this over a century ago:

"I kept six honest serving men, they taught me all I know;
Their names are what and why and when
And how and where and who."

These are the six key words for open-ended questions:

What
Why
When
How
Where
Who

They can be preceded by an introductory statement as indicated in the question above, or as shown below.

"I understand price is a key criteria for your selection, but if all the prices were the same, what would your key decision criteria be?"

There are two additional, stealth methods of asking open-ended questions, encouraging the customer to talk:

"Tell me about…"

"Describe for me…"

That's it. The salesperson who masters these six magic words and two phrases will have begun to master **Objective Based Selling**, and will have more professional skills and techniques than most of their competitors.

This book will supply over 100 open-ended questions specific to the sale of material handling equipment. These questions can be modified with word substitutions for specific material handling areas or products; and for specific situations.

Appendix 1 gives a summary of the most effective open-ended questions, organized by the stages of the **Objective Based Selling** model, as outlined on the **Objective Based Selling** diagram.

Funnel theory of questions

An effective method of using open-ended questions is to ask the broadest question first; then, based on the customer's answer, move to narrower questions. The benefit is, this gets the customer thinking widely about the project, so the salesperson has a better chance of learning all pertinent information. It also helps the salesperson avoid jumping to conclusions as to what is most important to the customer.

Example: the time question

Every viable material handling project has a time objective. Salespeople often jump to the conclusion that this objective is delivery-related, so they often ask: "When do you need it?" or some variation of that question.

Many salespeople don't even ask; they simply offer their delivery information: "Our delivery is running six months." "We have one in stock if you act quickly."

However, the customer's time objective may not be related to delivery. Other time-related issues in material handling purchases and projects are:

- Report due dates to the boss
- Budget deadlines
- Fiscal year issues
- Invoicing deadlines (also often tied to budgeting or fiscal years)
- Customer contacts leaving town and the need to get a project "off their desk" before leaving
- Get a project decided before a contact moves to a new job
- Construction schedules

And so on.

A broad, effective opening question about time is:

"What is the time frame for this project?"

Or, even broader:

"What is our time frame?"

Let the customer tell you what's important about time.

That can be followed up with narrower questions as the process proceeds, such as:

"What is the time frame for project review?"
"When do you need this information?"
"When do you need this proposal?"
"When does this need to be operational?"
"When do we need delivery?"

And, whenever a customer states a specific time frame is important, the follow-up question should be:

"Why is that (deadline) (date) (time frame) important?"

Similarly, in the money question, salespeople like to ask:

"What's your budget?"

Or the ever popular (and ineffective):

"How much do you have to spend?"

The broadest, first money question should be:

"What are the financial considerations in this project?"

Let the customer determine—and tell you—what's important about money.

Follow the trail of the answers

While mastering the open-ended questions to ask, salespeople should also learn to listen carefully to customers' answers, then follow the "trail of the answers." Based on the answer just given, what is the next most effective question to ask?

Example of a typical trail of answers (and follow-up questions):

"What are the financial considerations in this project?"

Customer's initial answer:

"I need to fight for budget approval from our Finance Committee."

This can lead to a whole set of follow-up questions, following the trail of the customer answers. Here are potential follow-up questions, assuming the customer has answered the earlier questions (this example does not show the customer's answers):

"Who is on the Finance Committee?""

"What is their role?""

"Who do you present your recommendation to in this
    Committee?""

"When do you do this presentation?""

"What format do you use to present your recommendations?""

"What are their decision criteria for project approval?""

"What kinds of projects are you competing against?""

"If you get turned down, how do you meet your
    objectives in this area?""

Ask effective open-ended questions. Listen carefully to the
answers. Follow the trail of the answers with strong, logical, follow-
up, open-ended questions. Follow the trail of the answers.

Customer reactions to open-ended questions

A common question of salespeople about the process of persist-
ently asking open-ended questions is: "What will the customer reaction be?"

Salespeople are often afraid of alienating customers with prob-
ing questions, and the more questions they ask, the more nervous
they get that the customer will be alienated.

Of course, salespeople must constantly watch and be sensitive to
the customer's reaction to the questions. But most customers want
to be asked questions. They know they need help, and are suspect of
any salesperson making recommendations without asking about
their operations.

If the questions are asked in an honest way, as part of a professional sales process, customers seldom refuse to answer or get irri-
tated—within some boundary, of course. As a salesperson senses he
may have asked one question too many, he simply needs to change
to subject or end the conversation for now.

It's easy to say this in a book—even backed up by years of sales
experience. But the only way a salesperson can really become con-
vinced of the efficacy and acceptability of asking open-ended questions is to do it.

In the author's experience, "Ask, and they will answer."

What to do after asking the open-ended questions

The step-by-step procedure is:

■ Ask an open-ended question

■ Shut up

■ Give the customer an interested look

■ Be quiet

■ Actively listen

■ Take notes as appropriate

■ Repeat as needed

Salespeople often find the second instruction on the list the most difficult to follow. When an appropriate, effective open-ended question is asked, customers often need a moment to reflect on their answer. Silence during conversation makes many people uncomfortable. As the silence lengthens, salespeople get nervous and actually sometimes jump to another question before their first one is answered. Or, amazingly, try to answer the question for the customer:

"When do you need this equipment?" ... "As soon as possible?"

A phrase sometimes used in communication is "pregnant pause." This is that time in a conversation when, after a question or statement, it's obvious something else needs to be said, but it's momentarily silent. Salespeople must learn to *continue* the silence and let the customer deliver the next statement; let the customer break the pregnant pause. While the pregnant pause may seem like forever, it actually seldom lasts more than 20 seconds.

Any salesperson who can't shut up for at least 20 seconds should find a new profession!

Chapter Eight

The Objective Based Selling Diagram— Your Memory Tool

The basic concept of **Objective Based Selling** is that people buying material handling equipment for companies make selections to satisfy:

- Business objectives
- Personal objectives

Specific **business objectives** for purchases or projects are often the result of internal negotiations among multiple decision influencers, including time and money considerations as well as many others specific to the given situation. Business objectives typical in material handling include: storing more in a given space; reducing maintenance costs; improving customer service; more accurate order picking; better inventory control; improving safety; improving material flow; reducing labor costs; shortening the supply chain; replacing equipment when lease periods are complete; and so on.

Personal objectives, on the other hand, are individual for each decision influencer. Personal objectives may be organization-oriented—objectives involved in advancing an individual's position in a company, or their career. They may help an individual meet their social needs. Or they may be more basic, involved with an individual's self-image, ego, or other personal factors.

The chapter immediately following this will explore some of these personal objectives commonly encountered in business-to-business selling situations. Subsequent chapters will deal with the process of uncovering business objectives and helping decision influencers negotiate those objectives within their organization. This chapter reviews the **Objective Based Selling** diagram shown on the next two pages.

Memory tool

The **Objective Based Selling** diagram is a memory tool to help the material handling salesperson remain aware of the two sets of objectives and of the individual, selling, and business' processes most effective in uncovering these objectives. It also helps the salesperson show the customer how they can meet their objectives by acting on the salesperson's customer-focused proposal.

The business and personal sides of the loop

The diagram is in the basic form of a loop, beginning with *Prospecting* (finding appropriate companies and individuals to engage) and ending in a *Commitment* (an order, a commitment to act on the proposal of the salesperson). The model indicates two parallel paths to a commitment—one based on customer business objectives and one based on customer personal objectives. All the phrases in the **Objective Based Selling** diagram boxes begin with verbs. Selling is a proactive activity.

For purposes of convenience, we will refer to these boxes as being on the business side of the loop or on the personal side of the loop.

Business side of the loop

On the business side of the loop, the steps from "Prospecting" to "Commitment" are defined by normal business practices. This book

OBJECTIVE BA

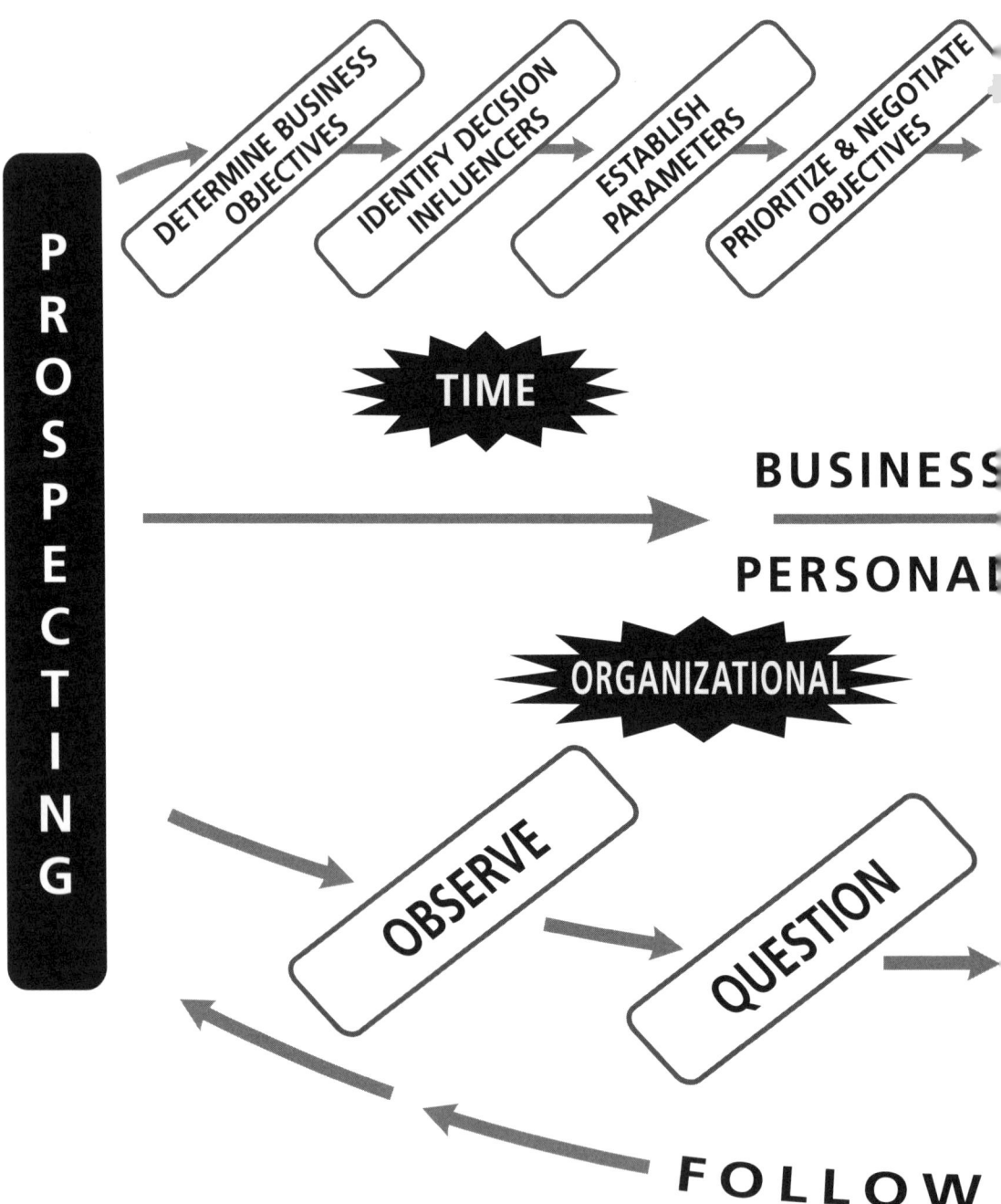

;ED SELLING™

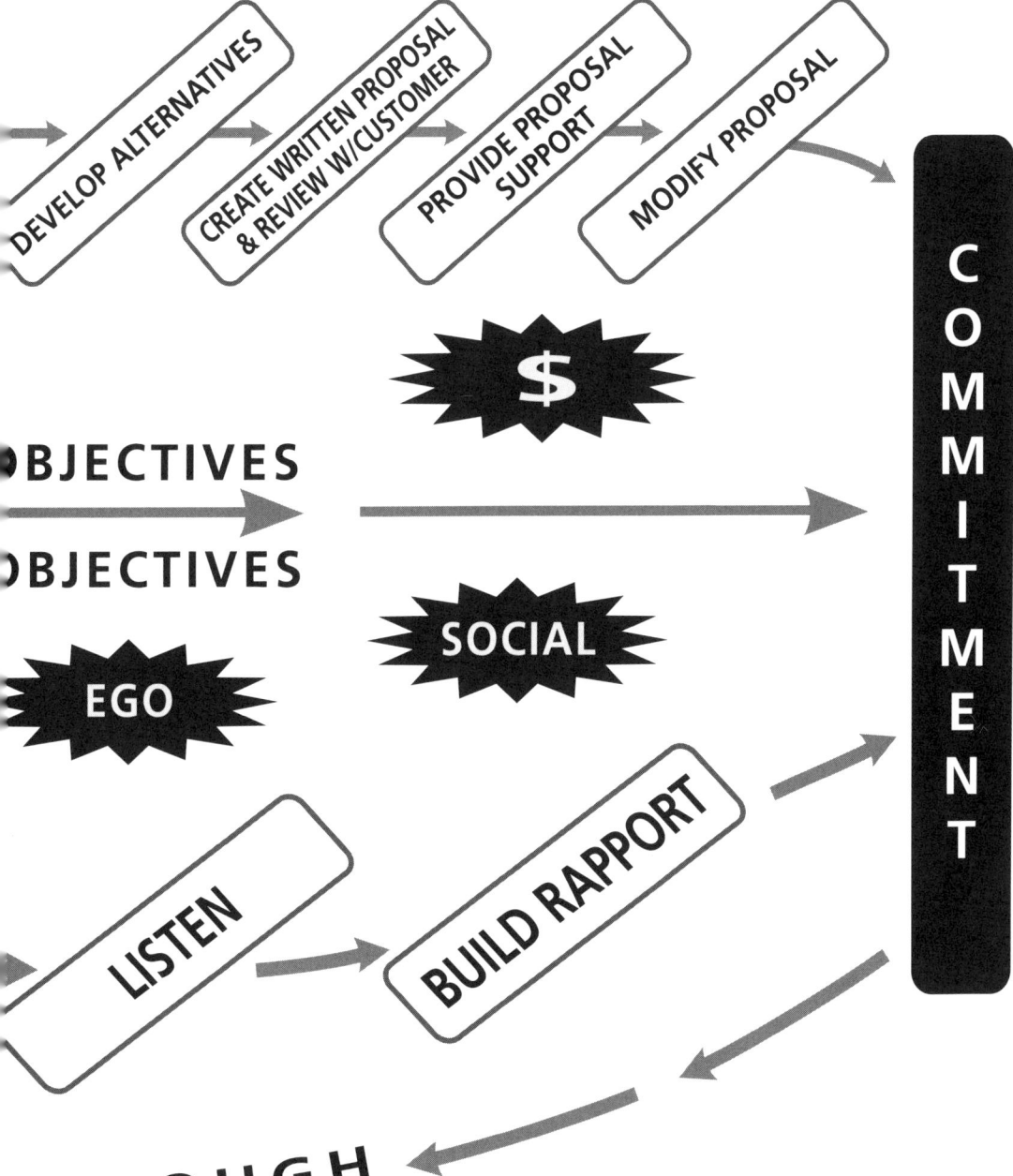

DEVELOP ALTERNATIVES

CREATE WRITTEN PROPOSAL & REVIEW W/CUSTOMER

PROVIDE PROPOSAL SUPPORT

MODIFY PROPOSAL

$

OBJECTIVES

OBJECTIVES

SOCIAL

EGO

COMMITMENT

LISTEN

BUILD RAPPORT

THROUGH

will devote one or more chapters to each step indicated on the business side of the diagram. This portion of the diagram outlines the basic steps in: uncovering business objectives, decision influencers, and parameters; helping customers negotiate objectives and consider alternatives; preparing and reviewing customer-focused proposals; supporting those proposals and modifying as appropriate to obtain a commitment—an order!

Personal side of the loop

This portion of the diagram indicates the less specific process of uncovering and working with customers' personal objectives: the techniques and skills of observing, questioning, listening, and building rapport.

Business columnists Sefflinger and Lubbers stated the following in their "Small Business Focus" column in the *Denver Post:* "Customers buy only two things: good feelings and solutions to problems." In the **Objective Based Selling** diagram,

"Good feelings" equate to meeting personal objectives.

"Solutions to problems" equate to meeting business objectives.

As the diagram indicates, progress is generally made on both the personal and business loops simultaneously. However, the diagram cannot indicate that progress is not always made at the same pace on each loop. In some situations, a salesperson almost immediately connects personally—instant rapport—but must work through the business processes successfully in order to get the commitment.

In other cases salespeople use navigation of the business loop processes to build rapport over time.

The diagram also indicates the importance of following through in material handling sales. The follow-through by the salesperson on the customer's commitment is the "rebooting" of the prospecting process. Successful follow-through should lead to another project with this customer, or allow the salesperson to obtain (create) a referral for a project with another individual or company.

The arrows on the diagram indicate the forward motion which is the responsibility of the salesperson to make happen. While not shown on the diagram, the tools the salesperson uses to maintain momentum in the sales process are open-ended questions.

On the business side of the loop, there are reminders of two objectives which are always there in "real" projects: time and money.

On the personal side of the loop, there are reminders that customer decision influencers have three types of personal objectives: organizational, social, and ego.

But every situation is different!

A common refrain of salespeople when considering the use of a sales model is, "Every situation is different."

And, indeed, every sales situation *is* different.

However, all business-to-business material handling sales situations have many, or all, of the characteristics of the material handling sales environment discussed in chapter three.

Use of the **Objective Based Selling** model and diagram is similar, for example, to diagramming football plays.

In football, every opponent is different; every stadium is unique; weather varies, even within the same game; personnel are constantly changing; every play develops differently. However, coaches know that all plays in similar situations have basic elements that are the same: the formations, blocking and tackling techniques, the importance of speed or deception, etc. So, while "Every football play is different," coaches keep diagramming and teaching plays with the basics so that individual players and teams have a reference point and can recognize the basic structure of each play as it develops, and use their specific skills to win. Having the play structure diagram in place frees the players to excel with their individual skills.

In **Objective Based Selling**, the open-ended questions, customer-focused proposals, listening skills, scrum meeting facilitation, etc. are analogous to blocking, tackling, reading defenses. The **Objective Based Selling** diagram is analogous to diagramming the play.

True, not every sales situation develops exactly as indicated in the diagram. In some situations the business side of the loop may be simplified or shortened due to clear objectives or decision criteria, fewer decision influencers, urgent time frames, better customer planning, smaller organizations and so on. In other situations the process may be extended or double back on itself as time goes by, as things change, as new decision influencers enter the picture, and as more complex, formal organizations are encountered.

On the personal side of the loop, where there are multiple decision influencers, personal objectives may be more complex, combining social, organizational, and ego-satisfying objectives of more than one person.

However, the common factor is that in all cases the salesperson must focus on the customer and let the customer tell them (in words and actions) how to sell them. The salesperson must do business the way the customer wants to do business. The **Objective Based Selling** diagram provides a road map to the process—the customer provides details in their specific situation.

It's up to the salesperson to recognize the variations of specific situations—to read the defense—and execute the basic skills necessary to focus on the customer and gain their commitment.

It's up to the salesperson to develop their individual skills and use them to win!

Readers are encouraged to refer to the diagram for reference while reading the rest of this book—and when selling! The format of the remainder of the book closely follows the **Objective Based Selling** diagram.

Chapter Nine

Prospecting

The verbal and visual image of the "49er prospector" panning for gold is often used for the sales activity of finding new customers, projects, or opportunities. It's an appropriate image. The prospector scoops up many sand grains in his pan, and sifts through them looking for a few nuggets of gold.

Similarly, the salesperson sifts through the many potential companies, contacts, and projects, looking for the ones with the most potential for the most profitable business within a reasonable period of time.

Rather than carry that analogy further, the appropriate question is: In the material handling, business-to-business sales environment, using the **Objective Based Selling** model:

What is sales prospecting?

Sales prospecting is a proactive activity aimed at generating a new "sales starting point." Specifically, prospecting is aimed at generating:

- New projects within an existing account
- New accounts with high potential for the salesperson's products or services in a significant amount
- Reactivation of "dead" accounts or projects
- Personal contacts with potential to lead to future business

OBJECTIVE BASED SELLING

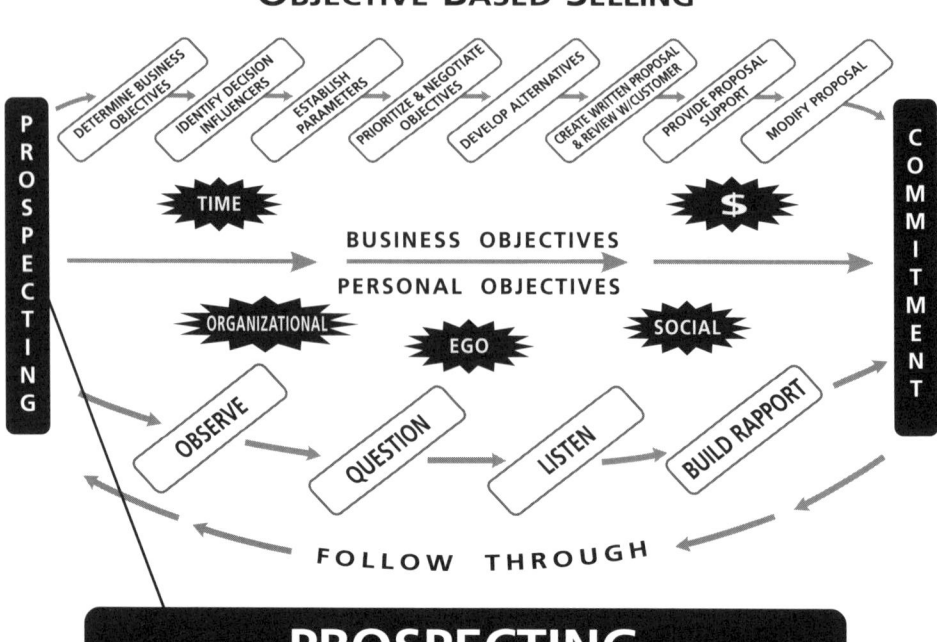

PROSPECTING

Sales prospecting includes qualifying. For a contact to be a viable prospect for a specific salesperson, there must be:

■ Personal engagement with an individual.

■ Realistic prospect of significant purchasing activity within 12 to 18 months.

■ Identified, real customer objectives.

■ Some sort of "match" between the prospect company and the salesperson's company. There should be compatibility in business practices.

■ Prospect company must have some openness to consider the salesperson's company as a possible supplier within a 12 to 18 month time frame.

The time frame of 12 to 18 months stipulated above, while somewhat arbitrary, encompasses the capital budgeting cycle of most companies and is a time frame over which a commissioned material handling salesperson can justify focused effort leading to a sale. Where projects are estimated to have longer time frames of activity leading to a sale, material handling salespeople must look to their company expectations, compensation plan, and other indicators as to how much effort and focus to commit. Most successful material handling salespeople will have a limited number of longer term projects with effort being managed through database and customer relationship management (CRM) techniques—in conjunction with guidance and sales management direction from their company.

This chapter will address the following sales prospecting topics:

- Who to target
- How to locate the targets: Techniques for making connections and qualifying
- Conducting an effective initial appointment

Who to target

Specifics of target companies for material handling prospecting depend on the exact product lines being offered and the marketing programs, customer base, and geographical coverage of the salesperson's employer. Some common identifiers of high potential prospects for material handling equipment and services include:

- Companies manufacturing physical products
- Warehouses
- Distribution facilities
- Service and repair facilities
- Building materials manufacturers and distributors with outdoor facilities
- Retail operations configured like warehouses
- Retail operations with large "back rooms" and distribution facilities to service them
- Construction companies (many material handling products are incorporated into the basic construction of all types of new or retrofit construction of all types of building facilities)
- Institutional facilities such as hospitals, penal institutions, universities, etc.
- Transportation and utility facilities such as rail, ship and container yards; airport and plane maintenance facilities; power plants, etc.
- Beverage production and bottling facilities

- Food processing and distribution facilities
- Maintenance departments of all types of facilities—
  including office buildings, apartment buildings, schools,
  universities
- State, local, and federal government facilities
- Department of Defense facilities—including on large navy
  ships
- Laboratories
- Archive record storage areas and dedicated facilities
- Printing facilities
- Professional maintenance companies (for equipment and
  facilities)

This is only a partial list. One of the positive things (from a sales standpoint) about material handling products and services is they are used in some form and quantity by almost all industries and commercial/institutional facilities. Even such an administrative-oriented industry like the securities industry has contracted for major automated material handling projects for literature and record storage, and document distribution.

One basic identifier of prospects for use and purchase of material handling equipment and services is *substantial physical facilities*. The primary purchasing and decision influencing persons may not be in the physical facility, but with diligence, they can be found by starting at these physical facilities and effectively using the open-ended questioning techniques of **Objective Based Selling**. The exact questions to ask are discussed later in this chapter.

Identifiers of high material handling potential include basic ongoing needs of large facilities, and add-on, modernization, and change in these facilities, particularly distribution and manufacturing facilities.

In addition to the list above, including substantial physical facilities, other qualifiers for high potential material handling prospects include companies experiencing:

- Growth
- Change
- Mergers, making acquisitions or being acquired

- Restructuring (even downsizing, which creates opportunities for helping with facility consolidation and space optimization)
- New or aggressive management focused on operations and facilities
- Start-up establishment of facilities
- Large contract awards and surges of business from their customers
- Construction programs
- Facility or equipment lease termination or renewal dates

The material handling salesperson should consider targeting:

- Customers his company is already doing business with
- Past customers no longer doing business with the salesperson's company
- Companies doing business with one part of the salesperson's company, or purchasing one part of the product line, but not others
- Accounts where company is performing service but not selling the new equipment
- Accounts where company is selling equipment but not doing the service
- National accounts of key company suppliers, which have facilities in the salesperson's area of responsibility
- Companies where significant personal, professional relationships exist with someone in the salesperson's company
- Companies being written about positively in the business news
- Large facilities
- Headquarters facilities (for regional, national, or international operations)
- Companies with identifiable seasonable needs
- Accounts where large proposals have been made in the past, but projects were dropped
- Accounts where large orders have been lost in the past
- Companies that commonly use equipment hard and "wear it out"

- ■ Competitors of companies where the salesperson and his company have completed successful projects
- ■ Key suppliers or customers of companies currently doing substantial business with the salesperson's company.

These last two are sometimes called moving upstream and downstream in the supply chain.

How to locate the targets: Techniques for making connections and qualifying

Again, the material handling salesperson should look to their company for the most effective prospecting techniques in their particular situation. Most companies in the material handling business have business development or prospecting techniques regularly, routinely, or periodically practiced. A responsibility—and opportunity—for material handling salespeople is to support and leverage these activities by their company.

A responsibility common in this area is to help build and maintain the company's (and salesperson's) sales and marketing database. Database software and use expectations in this are company-specific. Ongoing use of this database for marketing and business development is one way companies and salespeople maintain sales and marketing contact with companies and individuals who have high, but longer term sales potential—beyond the time window 12 to 18 months of primary potential.

Customer-focused prospecting

The prospecting techniques of many prospecting models are based on getting into a conversation with a customer and then quickly telling the customer all about the salesperson's products and services. Considerable time is often spent refining the salesperson's story. In some cases it is made very succinct, sometimes called an "elevator speech." This refers to what might be said about the salesperson's company or products to someone in the brief time frame of an elevator ride. In other cases, the salesperson and his company's story is expanded into a PowerPoint presentation or video or major presentation.

The problem with all these techniques is that they are focused on the salesperson, his products, and his company. They require the customer to passively listen while the salesperson lectures the customer.

"Here's what we can do and how we do it and how good we are at it and who we've done it for before and how good I am in following up and how long we've been in business and the broad product line we have and all the services we offer. Did I mention how good we are?"

At the prospecting stage—in fact at most stages of the sales process—customers are just not very interested in this story as normally presented, whether in an elevator speech or major presentation.

What the customer is really interested in is:

"What can you really do for me right now in my particular situation?"

In other words, the customer is more interested in himself and his company's situation or project, than in the salesperson and his company.

Of course, at the prospecting stage of the customer contact, the salesperson doesn't really know what, if anything, he can really do for the customer. At this point, the salesperson should be focused on the customer and learning about him, his issues, his decision-making process, and whatever else pertains to the situation.

Getting the customer talking about all this engages the customer in a conversation, instead of passive listening. The salesperson can then listen for a point of engagement where he and his company can be relevant to the customer's situation—where they can help the customer and his company meet their personal and business objectives.

This is most effectively done with open-ended questions and listening. So, in all the prospecting techniques discussed in the remainder of this chapter, the primary technique will be asking specific open-ended questions that focus on the customer, not the salesperson. This will lead to more meaningful conversations with customers, leading to more high potential sales opportunities. This will be particularly illustrated in a customer-focused format for initial interviews with customers.

Salesperson's primary prospecting responsibilities

In addition to supporting their company's business development efforts, material handling salespeople have primary responsibility for certain types of sales prospecting. These include:

1. Developing additional projects and more business with accounts already doing business with the salesperson or his company

2. Creating referrals

3. Reactivating lost accounts or projects

4. Developing local business with companies that the salesperson's suppliers have been successful with in other areas

5. Proactive prospecting with firms of high potential in their area of responsibility (however that area of responsibility is defined by their company—geographically, by industry, by assignment, etc.)

1. Developing more business with accounts already doing business with salesperson and his company

Here are questions which should continually be asked of all decision influencers the salesperson comes in contact with at accounts he is already involved with:

"What's changing in your (_____) operations?"

"What's your worst (forklift) (order processing) (loading dock) (maintenance) (conveyor) (storage) (add your own words) issue?"

"If you could change one thing about your (_____) operations, what would it be?"

"What one thing, if you could do it, would dramatically, positively change your business?"

"Where do you need more space?"

"What's your biggest maintenance issue with (_____)?"

"What capital projects are budgeted for this year? For next year?"

"What uncommon or special circumstances cause you problems?"

**"How satisfied are you with your current supplier of
(_____)?"**
**"What dissatisfaction do you have with your current
supplier of (_____)?"**

Salespeople are encouraged to customize these basic questions to
their specific products and services.

2. Creating referrals

When most salespeople are asked the source of their best sales
leads, almost invariably they answer "referrals." That is, recommendation from a customer with whom the salesperson is already doing
business to others with similar needs.

Of course, many referrals take place without the salesperson's
knowledge, in conversations which may happen in or out of the business environment. The salesperson learns about them when the new
prospect contacts them and says something like "I was talking with
Harry at XYZ Company and he's very satisfied with work you've
done for him." Wow.

When working with sizeable companies, the conversation and
referral may take place within the current customer's company—in
another area or department or location that the salesperson was not
familiar with or not doing work for.

Being the beneficiary of these types of referrals is not a proactive
activity by a salesperson. These are essentially uncontrolled, serendip-
itous results of good work done for a customer. However, salespeople
can create referrals with the right attitude and questions.

The attitude is that referrals can be created; salespeople don't
have to "hope" they will happen. Hope is not a strategy or proactive
activity.

To create referrals within a current customer, the questions are:

**"Who else should I talk to about (material handling)
(space optimization) (cost reduction) (_____) projects
in your company?"**
**"If you were me, who else would you want to talk to in
your company?"**
"Who in your company gets involved in (_____)?"

"Who in your company has responsibility for (_____)?"
"What other (locations) (departments) (divisions) have
    this issue?"

To create referrals outside the company of the salesperson's current contact, the questions are:

"Who are your customers?"
"Who are your suppliers?"
"Who are your competitors?"
"Who do you know in this industry that's growing?"
"Who else in this business park do you know?"
"What professional, industry organizations do you
    belong to?"

(The idea here is to get names of others in similar positions in
other companies, perhaps by purchasing a list or joining the organization yourself, or simply using this to trigger the idea of other
names from your contact.)

"Who do you know that could use my help?"
"What industry publications do you read?"

(The idea here is to subscribe to those publications yourself, to
stay current in the customer's industry, or perhaps purchase a list
of subscribers.)

"Who can you refer me to?"

When do you ask these questions? All the time. Some companies
have an expectation of their salespeople that they will gain some
referral in every customer meeting or contact. Referral questions
can even be asked when prospecting unsuccessfully with a contact.

"I understand you don't have a current need. Who do you
    know that might have a current need?"

Referrals can be created with the right questions.

3. Reactivating lost accounts or projects

It's all in the questions.
Make the contact, and ask:

"Last year we were unsuccessful on our proposal for
    your (_____). What would you have liked to see your
    selected supplier do better on that project?"

"Prior to last year, we were your primary supplier for
(_____). What is your level of satisfaction with your
current supplier?"

"What would you like to see your current supplier do
better?"

"What's changed since then?"

"What new projects are you planning for this year?"

"What areas are you focused on this year?"

Once the conversation is initiated, many of the questions listed under "Developing additional projects and more business with accounts already doing business with" (number 1 above) can be used.

4. Developing local business in companies where salesperson's suppliers have been successful in other areas, or where suppliers have national accounts

Basic information on the application and success of supplier products or agreements in other areas is usually easily obtainable from the suppliers or the salesperson's company. This information provides a natural first point of local contact:

"We're the local supplier of (_____), which has been
used successfully in your facility in (_____). I'd
welcome the opportunity to learn more about
your local operations to help determine if this
would be beneficial here. Who should I meet with
to discuss this?"

The variation on this approach for national agreements or con-tracts is:

"We're the local representative for (_____), which has a
national agreement with your company. Who should I
talk to, to determine if you are getting all the local
benefits of this agreement?"

Notice in both cases the focus of the question is the benefit to the customer of scheduling a meeting.

5. Other prospecting techniques and questions

When following leads (indications of interest or requests for information) generated by company or supplier marketing efforts such as websites, advertising, yellow pages, etc.

After a brief introductory statement such as:

"Hi, I'm Gary Moore with XYZ Company. I've been informed you recently requested information on (_____)."

> **"What application did you have in mind when you requested that information?"**
> **"What triggered your interest in (_____)?"**

When participating in a product or trade show

When a prospect specifically stops at the booth and appears to have interest, ask:

> **"What triggered your interest in (_____)?"**

When working to engage more casual booth visitors (or even aisle strollers!), ask:

> **"What brought you down to the show?"**

Follow-up questions can be:

> **"What's your area of responsibility?"**
> **"What projects are you currently involved in?"**
> **"What's changing at your company? In your area?"**
> **"What is the best way and time to contact you after the show?"**

Most likely, a sale cannot be made at a trade show, unless it is a planned booth visit regarding an ongoing project. The primary purpose of the show is to identify prospects with projects and get their permission to contact them for an appointment after the show. The questions listed above will do that.

Gaining control of incoming calls

Material handling salespeople often "field" incoming calls from customers or prospects asking for information. The customer is leading with the question, and the natural reaction of most salespeople is to give answers. A better strategy is to gain control of the call by acknowledging the customer's question and then asking open-ended questions before giving answers.

Slow the customer down by asking some simple questions first:

"Thanks for your call. I'm sorry, I didn't get your company name. What is your company? What is your full name? What is your area of responsibility?"

Of course, it is important to take notes!

"What triggered your call for information on (_____)?"
"How did you choose to call us?"

Before recommending or pricing specific equipment, there are questions to be asked about specific details and parameters of the customer's specific application and situation.

At the end of the call, ask:

"What's the best way to contact you in the future?"

Cold call prospecting "in the field"

In the assigned areas for some material handling sales territories, it may still be possible and appropriate for salespeople to literally go "door-to-door" physically calling on industrial or commercial facilities to prospect. At one time this was a primary prospecting method. In some instances it developed the nickname "smokestacking." This came from the fact that most users of material handling equipment were manufacturing facilities whose facilities literally had smokestacks.

The primary purpose of this type of cold calling in the field now is to get information for database additions, or future mailings or teleprospecting. In the 21st century this type of prospecting is usually only possible or most effective in medium and smaller size, locally owned companies. However, it can be tried with any facility where access is available. For larger facilities, national firms, or firms with high security arrangements, other methods must be used. These will be discussed in following sections of this chapter.

Questions to be asked of receptionists or whoever is available during "cold calling" are:

"Who should we contact in the future about warehouse
   equipment? Material handling equipment? Needs in
   the areas of (forklifts) (loading dock equipment)
   (casters) (_____)?"
"Who in your facility gets involved with (warehouse
   equipment) (material handling equipment?)
   (forklifts?) (_____)?"
"What's the best way / time to contact him?"

Ask for a business card with contact information. Move on to the
next facility. Follow up later as appropriate.

A recently developed 21st century approach to "smokestacking"
involves reviewing satellite photos of areas, looking for warehouse
facilities, usually identified by their roofs! This involves subscribing
to one of the aerial photo services.

Letter sets up phone call

An aggressive, proactive sales prospecting technique which goes
beyond simply asking questions of current contacts is called "letter
sets up phone call" prospecting. Using information from company or
purchased databases, or from "smokestacking" or other sources,
salespeople send letters to prospects which give a brief overview of
objectives they accomplish for their customers. These letters state
that the salesperson will be calling to learn more about the
prospect's company and operations and, if appropriate, set an
appointment to learn more, to see if they can be of help to the
prospect.

The differences between this and many prospecting letters and
techniques include:

- Letter does not attempt to "introduce" the salesperson's
  company; it instead talks of business objectives it helps
  customers with. It's a customer-focused letter, not a "me"-
  focused letter.

- It doesn't say too much; it simply introduces ideas and "sets
  up a phone call by the salesperson." Most of the information
  is in "bullet form," not long paragraphs that the customer
  won't read.

■ Instead of instructing the customer to "call if you need anything" (a common, ineffective approach in sales letters and in sales conversations), it indicates the salesperson will do the calling.

Email is certainly a more common form of communication today than written paper letters. That is one reason a paper letter is encouraged in this type of prospecting—it will stand out. Many emails today are deleted before even being opened. If desired, a small brochure and business card can be sent with the paper letter. A personal signature at the bottom of the letter gives it a personal touch, perhaps a first step in building a personal, professional relationship.

Sample prospecting letter that "sets up phone call"

(Inside Address)

Attn:

Dear _____:

As (material handling specialists), we help companies accomplish (warehousing) objectives such as:

- Storing more in less space
- Being compliant with OSHA operator training requirements
- Reducing (forklift) (conveyor) (loading dock) (door) maintenance costs
- Improving order picking accuracy
- Streamlining material flow
- Increasing productivity of material movement operations
- Lowering energy costs
- Better inventory control
- Designing ergonomic work stations
- Safer operations, fewer injuries, lower insurance costs

The enclosed brochure indicates some specific areas we are involved in. I will call within the next two weeks to schedule an appointment to learn more about your operations and share some ideas with you.

Sincerely
Salesperson

P.S. Our website, www.xyz.com, has some informational resources on the above topics.

(Of course, the bullet points should be customized to your company and prospecting target. There should be no more than four bullet points. Enclose a short brochure and business card. Make that follow-up call!)

Word processing capabilities make the customizing and mail merging of such letters systematized and low-effort.

Of course, the second part of the "letter sets up the phone call" prospecting process is the *phone call*.

These should take place approximately a week to ten days following the letter. While many decision influencers use voice mail as a block to incoming calls, with persistence, most can be reached. There are several strategies: calling early, before the official "business day" begins; at lunch; or late in the day. In some cases, Saturday morning or other weekend call strategies can reach busy decision influencers who are working long weeks.

Another strategy for getting through is to enlist gatekeepers' help. Call the general company number and ask the receptionist, or other gatekeeper, for help in reaching an individual. This may be someone the salesperson has met on a face-to-face smokestacking call.

If, after five tries (this is a good rule of thumb maximum, unless the prospect is of such high potential that more tries are appropriate) all that's reached is voice mail, a voice mail message can be left, with a follow-up letter. It is recommended no voice mail message be left until the salesperson is ready to move on to another prospect.

Here are three alternate "scripts" for the follow-up phone calls, when the prospect is reached:

Brief introductory statement:
"This is Gary Moore with XYZ Company. I recently dropped you a note regarding material handling objectives I've helped several companies achieve in their warehouse and distribution operations."

Then,

Script alternate one
"What objectives are you focused on in your warehousing operations?"

Script alternate two
"I'd like to schedule an appointment to learn more about your operation and where our capabilities might help you meet your objectives. What's an appropriate time?"

Script alternate three

"Who should I schedule an appointment with to learn more about your operations to determine where we might help you meet your objectives?"

The concept of this technique is not to try to "tell" the customer a lot of things on the phone call, but to get them talking and to go for the appointment as soon as possible if it seems appropriate.

If the appointment is not feasible on this call, ask some further open-ended questions to trigger more interest, or to make any follow-up mailing targeted. Questions such as:

"What's changing in your operations?"

"What are your improvement objectives for the coming year?"

"If you could change one thing about your warehousing operations, what would it be?"

Based on responses to these and other questions, the salesperson can determine what is appropriate for a follow-up mailing, or how aggressively to pursue for an appointment.

Voice mail script

In the situation where the salesperson simply cannot connect by phone with the prospect, after approximately five tries, a voice mail can be left. An appropriate voice mail is:

"Mr. Trent, my name is Gary Moore. A couple of weeks ago I dropped you a note indicating objectives we've helped customers achieve in their material handling and warehousing operations. I apologize for being unable to connect by phone. I would like to meet you and learn more about your operations, to see if we could also help you meet your warehousing objectives. I will drop you another note with further information, along with my business card. I can be reached at my cell phone number xxx-xxxx; my email address will be on the card. Thanks!"

Using a welcoming voice which does not sound like it's reading a script is important.

Whatever is mailed should include a postage paid return mail card along with the salesperson's business card. By giving the cell phone number in the voice mail, the salesperson is indicating his trust in the prospect, and his sincere intention to connect.

Conducting an effective initial appointment

The prospecting techniques described in this chapter all have the goal of an initial appointment with a prospect. This should not be termed an "introductory appointment." That implies the salesperson has the appointment to "introduce their company, products, and services." Instead, it should be an "initial appointment" (implying there may be more) with the goal of learning about the customer, their area of responsibility, their objectives.

A quality "initial appointment" with a prospect is:

Five to twenty-five uninterrupted minutes with a decision influencer to learn about their operations (including business and personal objectives) with a goal of uncovering how the salesperson can now, or later, help them improve their operations—meet their objectives.

Before the appointment, it's good strategy for a salesperson to do some research on the target company. This can include visiting their website; checking company records to determine what business the salesperson's company has or is doing with the prospect; checking key suppliers' national account agreements; business or news websites which the salesperson's company may subscribe to, such as www.hoovers.com; or even "Googling" the company (or using the search engine of your choice, of course!).

Because material handling salespeople are dealing with physical things—forklifts, conveyors, casters, loading dock equipment, shelving, racks, palletizers, vertical lifts—the tendency is to bring all sorts of catalogs on the initial call, including the ubiquitous "company brochure" so the salesperson can show the customer.

Wrong strategy. That is product-focused, not customer-focused.

The material handling salesperson should consider going on the initial appointment with prospects with just their pad, their questions, and their attention.

After initial social courtesies—kept to a minimum—the salesperson can begin.

"Thanks for taking time to meet with me. As specialists in
(material handling) operations, we have helped many
customers achieve objectives such as (storing more in
less space); (reducing forklift operating costs); (improving safety at the loading dock); what areas of your
warehousing operations are you working to improve?"

As much as possible, the rest of the interview should center on the
customer and his operations, not the salesperson and his company.

Depending on the specific product lines being sold by the salesperson, many of the questions coming up in chapters ten and eleven
will be appropriate here. A few examples include:

"What's your area of responsibility?"
"What is your biggest (forklift) (conveyor) (order picking)
(space) (safety) (loading dock) (_____) issue?"
"If you could change one thing in your (_____)
operations, what would it be?"
"What is your worst (forklift) (space) (_____) issue?
"What material handling projects are you involved in
this year?"
"Where are you looking to improve?"
"Tell me about your material handling operations."

As customers begin to answer these and other questions, they
often will "open up" about their operations and tell more than was
even asked. The salesperson can follow the trail of the answers!

If appropriate for the time frame, situation, and customer, it is
always good to get the opportunity to view the customer's operations.

"What would be an appropriate time to take a look at
your warehousing operations?"
"What's the possibility of touring your warehouse?"

These opportunities give the chance to learn more about the customer operations and to begin developing a relationship. It can also
be an opportunity to meet other decision influencers.

At a point in this initial appointment it will most likely be appropriate for the salesperson to briefly tell about his company. A good
place to start is with these questions:

"What has been your experience with our company?"
"What is your impression of our company?"

"**What would you like to know about our company?**"

It may also be appropriate to learn about the competition:

"**Who are your current suppliers for (forklifts) (conveyor) (order picking systems) (shipping or receiving equipment) (_____)?**"
"**What do you like about them?**"
"**What would you like them to do better?**"

In all these initial conversations, the salesperson is looking for ways he and his company can benefit the customer. One goal of the initial appointment is to schedule a follow-up appointment: *time with a decision influencer to develop appropriate ideas, advance projects, or respond to stated needs as learned in the initial appointment.*

Prospecting summary

This chapter has introduced several prospecting techniques for material handling salespeople. There are many others; many material handling sales companies have systematized this process as business development or marketing.

The key **Objective Based Selling** prospecting concept is:

Focus prospecting techniques on the customer—on learning what they are focused on, by using various open-ended questions, searching for a project or selling opportunity where the salesperson can help the customer meet their objectives.

The goal of prospecting is to schedule an initial appointment with a high potential user or purchaser of material handling equipment.

The initial appointment should be customer-focused, looking for customer objectives the salesperson can help achieve.

And, make more calls

Everything else being equal, the salesperson who makes the most calls will make the most money. A certain element of selling remains a numbers game—the more sales calls made, the more initial appointments, the more facility tours, the more proposal opportunities, the more proposal reviews... the more sales. The use of **Objective Based Selling** techniques makes all these actions along the way more effective, but work ethic still counts!

Chapter Ten

Personal Objectives

Before beginning a discussion of the sales processes for uncovering business objectives on the business side of the loop (in the next chapter), it will be helpful to have a brief discussion of personal objectives often encountered on the personal side of the loop.

Personal objectives of decision influencers in material handling sales generally fall into three categories:

Organizational (getting ahead or getting by in their own organization; building their career)

Social (doing business with people they enjoy doing business with, in a manner they enjoy)

Ego-centered (dealing with or satisfying their self-concept)

More specific personal objectives in each category include:

Organizational personal objectives
- Look good for the boss
- Gain more responsibility
- Earn a bonus
- Get a promotion
- Meet organizational criteria

- Get organizational attention
- Don't rock the boat!
- Have a salesperson do my work for me (data gathering, communication, etc.)
- Add to my career resumé

Even business owners have personal organizational objectives:

- Look good for subordinates
- Set organizational tone
- Avoid risk to the organization

Social personal objectives

- Do business with people I like to do business with
- Don't do business with people I don't like
- Spend time with a supplier who shares a common interest
- Pay back a favor to a supplier or supplier representative
- Reward effort (for example, data gathering, multiple proposal modifications, exceptional follow-through)
- Play more golf!

Ego-driven personal objectives

- Be reassured
- Be on the leading edge
- Display business competence (look smart!)
- Be professional (meet external or self-imposed professional standards)
- Appear to be in charge
- Hide a feeling of inferiority
- Display a feeling of superiority
- Be in control

The "Get it off my desk" personal objective

There is one personal objective which is encountered more often than might seem likely: It's called the "Get it off my desk" objective. This is commonly encountered when a primary decision influencer—perhaps a project champion—is facing extended time away from the office for a business or personal reason. It may be a vacation; an extended business trip; dealing with a personal issue. When this occurs near the culmination of a project or project decision process, there is a tendency to either postpone the decision until after the time away, or speed up the decision to "get it off his desk" (and mind) before he leaves. More often than not, the latter is the case.

When a salesperson is aware of these situations and senses the personal objective of the decision influencer to make a decision—and purchase—before this often self-imposed deadline, it will be in the salesperson's best interests to stay in close contact with that decision influencer; give them high levels of access to the salesperson (personal cell phone numbers for example), and be in a position to respond quickly to requests for information, proposals, or modified proposals. Helping a customer "get it off their desk" before a time away can be a powerful way to help customers meet an objective—and to *obtain commitment* (get an order!).

More later

This chapter was meant as a brief introduction to the personal side of the **Objective Based Selling** loop. The next several chapters will focus on the business side of the loop. Later, discussion will return to effective methods of uncovering and understanding the personal objectives of customers, and how to help them meet those objectives.

Chapter Eleven

Determine Business Objectives

The job of the salesperson in **Objective Based Selling** is to determine the customer's business and personal objectives, and show how the customer can meet their objectives by acting on the salesperson's proposal.

Once a potential customer project or purchase intent is identified, the salesperson should begin the process of identifying the objectives of the project. In other words, why is the customer considering this project or purchase? What are they trying to accomplish?

The simplest questions to determine business objectives are variations of:

"What are your objectives in this project? In making this purchase?"

One benefit of using the word *project* is it elevates the situation to a more important status than simply "buying something." This can help your contact compete for funds within his company; it also gives your contact a higher level perspective on the situation. If he is simply making a purchase, price may be a more critical criterion. If he is implementing a project, other factors may dominate—like trust that the project can meet stated objectives and that the sales-

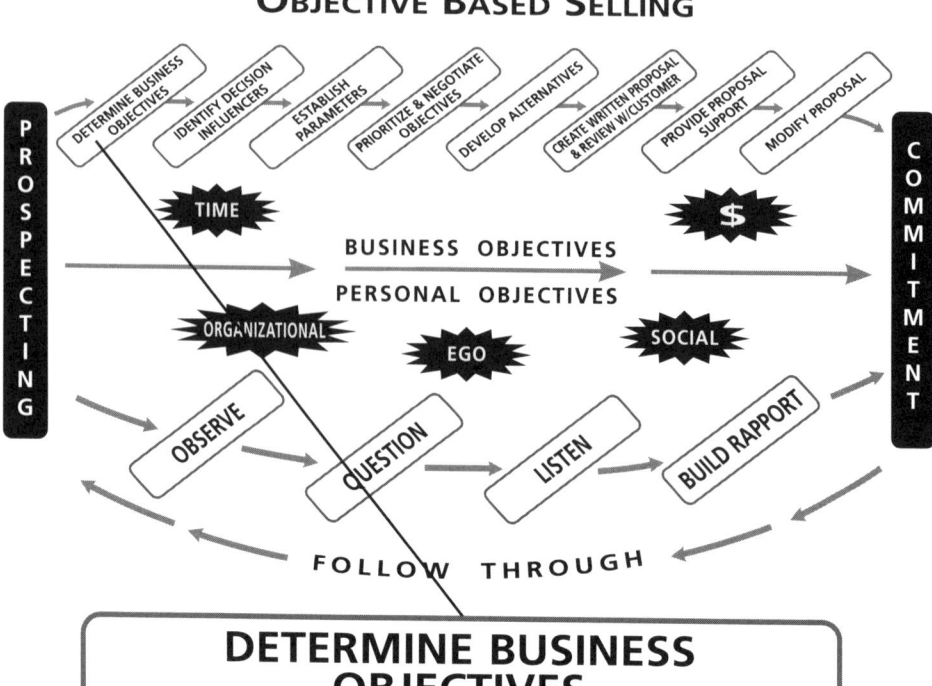

person will follow through during implementation.

Other versions of this "objective" question include:

"What are you trying to accomplish with this purchase?"
    "With this project?"
"What triggers this project?"
"What's the most important thing about this project?"

In the material handling sales environment, there are usually multiple decision influencers. These questions should be repeated for each decision influencer met. Where there are multiple objectives in a project or purchase; or where different decision influencers have different objectives, these will be negotiated and prioritized at a later phase in the sales process.

Also, over time, project objectives may change, so the question should be repeated at appropriate intervals.

Objectives stated by the customer should be written down, indicating their importance and that the salesperson is paying attention and is focused on the customer.

Time and money objectives

Every meaningful project with a real possibility of happening will have two sets of objectives always present (in addition to the others). These are time objectives and money objectives. If these are not volunteered in answers to the first, broadest objective questions (listed above), they must be identified with more specific questions.

The broadest time question is:

"What is your time frame on this project?" "With this purchase?"

The broadest money question is:

"What are your financial considerations in this project?" "With this purchase?"

Narrower questions can be added in pursuing the trail of the answers. For example, if the customer states a delivery time is the critical time frame, the follow-up questions might be:

"Why is that time frame critical?" (This question could uncover important additional information about decision influencers, other projects, and plans.)

"What happens if we miss that time frame?" "What happens if no one can meet that time frame?" (These questions help test how critical that time frame really is, and might also tell the salesperson how to handle it if he really can't meet the customer's stated time frame.)

If the answer to the money question is:

"Be under our budget."

The follow-up questions might be:

"What is that budget?" (Well, let's get the numbers out on the table!)

"How was that budget established?" (All budgets are a result of a negotiation. Find out how they do that.)

"Who determines the budget?" (Here is a chance to meet more decision influencers.)

"What does that budget include?" (For instance, freight? Taxes? Training? Often, these are handled outside of equipment budgets.)

"What do you do when project estimates are over budget?"
"How do you handle over-budget situations?" (Companies
make purchases outside of budget limits all the time.
Introduce the idea, and find out how it can be done.)
"How do you plan to finance this purchase?" (Maybe a
lease is possible, or payments, or trading in old equipment,
or through other means.)

Soft objectives versus hard objectives

Identifying objectives is also a qualifying phase of a sales situation. The more specific the objectives, the more likely the project or
purchase will really happen. Projects with hard objectives have been
thought through and discussed inside the customer's organization.
Soft objectives, on the other hand, may indicate the project is still in
the "thinking" phase and the company is not yet committed to it.
Understanding the customer's commitment and sense of urgency
about a project or purchase will help the salesperson determine how
much time and resources to devote to it at this time.

Examples of soft objectives might be:

"Improve safety"
"Better customer service"
"Reduce maintenance costs"
"Upgrade our fleet"
"Create more space"

Hard objectives, on the other hand, are more specific, usually
with numbers or visual images associated with them. Examples:

"Reduce back injuries in our palletizing area by 50 percent."
"Improve order accuracy from 98.2 percent to 99 percent."
"Increase throughput 35 percent in order to handle our new line
in the same facility with the same people count."
"Reduce forklift maintenance costs by 30 percent."
"Have our new order fulfillment line up and running by October
1, in order to handle holiday orders."

The more specific the objectives, the more the project is likely to
happen. And with specific, hard objectives, the salesperson and his

technical support can evaluate their capability to meet the customer objectives.

"Why are we doing this project at all?"

Many material handling salespeople begin responding to a customer's project or purchase request without ever truly understanding why the customer is considering the project. By forcing the customer to articulate this, using as specific terms as possible, the salesperson is performing a sales function—helping the customer understand how to justify the project internally—and compete for funds. If there are no specific, identifiable, hard objectives, the chances of the project ever really happening are greatly reduced.

Understanding the customer's business objectives is understanding: **"Why are we doing this project?"** (another question that can be used to uncover business objectives). And that's a long way toward being on the customer team that implements the project!

Repeating objectives back to customers

A key element of the effective written proposals of **Objective Based Selling** will be to repeat specific project or purchase objectives back to the customer, in writing, in the proposal. They will be presented as bullets, prioritized, following the statement:

"As we understand them, your objectives in this project are:"

There is power in showing these to the customer's multiple and hidden decision influencers at the proposal phase. It focuses the proposal on the customer instead of the "stuff." It helps show that your proposal is based on an understanding of the customer's situation; it builds trust; it causes the customer to pause and reconfirm these objectives—or recognize they have changed. This restatement of the customer objectives is also a proposal differentiator, because most competitive proposals won't have this.

Chapter Twelve

Identify Decision Influencers

As discussed earlier, material handling sales most often involve multiple, and hidden, decision influencers. The larger, more complex the sale, and the longer the time frame for a decision, the more decision influencers are likely to be involved.

Identifying, and where possible meeting and developing relationships with as many of these decision influencers as possible is important for the effectiveness of material handling salespeople. No person or job title in a customer organization should be overlooked as a possible decision influencer. Equipment operators, engineers, managers, administrators, superintendents… even receptionists may influence decisions.

There are several specific questions which should be used by salespeople to identify decision influencers and understand their role in the decision process. The broadest questions are:

"Who, besides yourself, is involved in this project?"
"Who, besides yourself, is involved in the decision-making process?"
"Who, besides yourself, will influence this decision?"
"What are the steps in your decision-making process?"

OBJECTIVE BASED SELLING

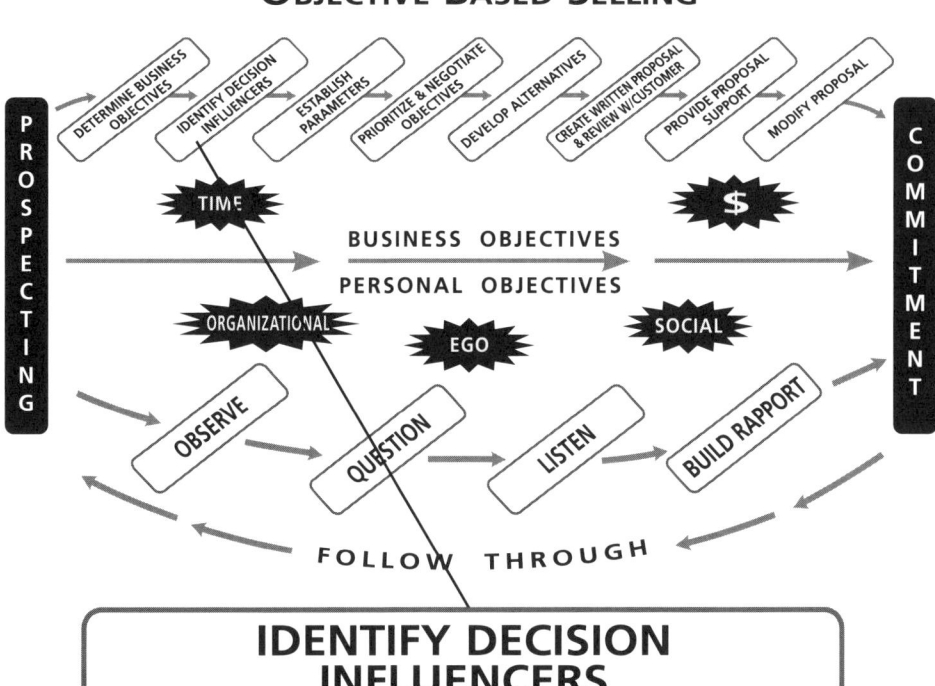

IDENTIFY DECISION INFLUENCERS

"If you were me, who else would you want to talk with about this project?"

"Who can give me the most details on your parameters, so my proposal can be more responsive?"

More narrow, specific questions to identify decision influencers draw on the material handling salesperson's knowledge that certain job responsibilities are routinely involved with many material handling purchases. Depending on the type of project, one or more of these or similar questions may be used:

"Who is responsible for safety in this operation?"

"Who works with the financial approvals on these type projects?"

"Who's in charge of your loading dock?"

"Who has responsibility for maintenance?"

"Who is the operational manager?"

"Who is involved in the (_____) aspect of this project?"

Follow-up questions can include:

"Who does that person report to?"
"Who has overall responsibility?"
"What is your role in the process? Project?"
"What is that person's role in the process? Project?"
"Who has the most say in this project?"
"Who is against the project?"
"Who is the strongest advocate for this project?"

Other, more specific methods of uncovering decision influencers may include asking the customer for a copy of their organization chart; or visiting their website and doing a little research.

The salesperson should look for every opportunity to meet any and all decision influencers identified. One way to accomplish this is to simply ask, **"What's the best way to meet him?"** or **"What's the possibility of meeting him?"**

As discussed earlier, material handling decision processes take time. Over time, the number, identity, and role of decision influencers can change. So, these questions, or variations, should be constantly repeated as the project progresses. One variation can be:

"What's changed about the decision process (or about the decision influencers) since we last discussed this?"

Job titles

Many salespeople find it natural in a business sales situation to ask a person they have just met, "What is your title?" This is not a good question to ask, for the following reasons:

- It implies the person's job title is important. It may or may not be.

- In fact, some job titles are somewhat demeaning—the person being asked may not be proud of his title.

- Similar titles indicate different things in different organizations, and are often misleading as to real responsibilities. Titles are not as significant to salespeople as the area of responsibility and level of influence. Note: Titles should not be totally ignored; they are important in formal communication or introductions to others, or potentially as indicators of importance in an organization. However, there are simple ways of getting the title of persons; their business card is the most obvious. Titles can also be obtained by asking receptionists, visiting websites, or asking others.

■ Material handling salespeople should be much more concerned about what their contacts' responsibilities are, and the role they play in the decision-making process. Titles don't clearly indicate these facts.

The material handling salesperson should replace "What is your title?" with the following, when meeting people at their customers and prospective customers:

"What is your area of responsibility?"

And the follow-up questions:

"What is your relationship to this project? Purchase? Area?"

And again:

"What is your role in the decision-making process?"

Chapter Thirteen

Establish Parameters

Because material handling involves equipment, along with installation, services, and controls of equipment, the overwhelming tendency of both customers and salespeople is to discuss specifications of equipment or controls early in the sales process. When specifications are discussed before appropriate questions are asked and investigations done regarding current and projected operational parameters, assumptions are usually made or implied. Often, the selection or recommendation of equipment specifications is based on things as simple as "What we've always used" or on statements like "Our heaviest loads are those boxes of forms—about 3,500 pounds per pallet, I think."

This often leads to costly mistakes, such as forklifts that won't fit through the door; conveyor that won't handle all the boxes or move fast enough to meet requirements; pallet rack without the capacity for the heaviest loads, or which has more capacity than needed (resulting in higher prices and lost orders); casters that wear out too fast, or won't roll easily; dock levelers that won't safely handle the lowest trailers; a failure of control software to communicate with customer software; and so on. The discussion of equipment specifications prior to completely understanding and defining the cus-

OBJECTIVE BASED SELLING

tomer job to be done is a violation of the key precept of **Objective Based Selling:** Focus on the customer.

Before discussing equipment in detail, material handling salespeople should first discuss parameters of the customer's operation and job to be done. In some sales models this is called fact finding, or the investigation phase. In **Objective Based Selling**, it's **Establish Parameters.**

What are parameters?

A general dictionary definition of parameters is:

> *"Any set of physical properties whose values determine the characteristics or behavior of something."*

The physical, numerical and interface properties of the customer's job to be done will determine the characteristics of the equipment, service, installation, or controls to be recommended.

The parameters appropriate for proper specification selection and recommendations of each particular type of material handling equipment vary.

For example, for pallet rack, key parameters include:

Load width, length, height, weight, and weight distribution; floor thickness, construction, and condition; clear ceiling height; loads to be stored per area; pallet condition; type of forklifts being used; susceptibility of rack to forklift impact; applicable fire regulations and building permit standards.

For loading dock equipment, key parameters include:

Highest and lowest trailers to be serviced, measured from driveway surface; maximum loads to be handled; types of forklifts to be used; load cycles; load dimensions; inside trailer dimensions; dock height from grade; driveway and approach grades; dock concrete specifications, type of material being handled, weather extremes.

The specific parameters to be investigated and addressed for each type of equipment are beyond the scope of this book. As mentioned earlier, these are to be included in the **product knowledge** aspect of the material handling salesperson's training.

But, repeating the basic premise: the material handling salesperson should focus first on the customer—on physical and numerical parameters of the job to be done and surrounding customer circumstances—before recommending or proposing equipment.

In addition to making fewer equipment specification mistakes, another positive aspect of this approach is that most customers understand it's in their best interests. This also helps progress on the personal side of the loop as well as business side of the loop. Customers trust salespeople who don't jump to conclusions but who "do the homework" required to help customers meet their objectives.

How to establish parameters

Usually, salespeople initially rely on customers to provide the parameters, or they must obtain permission and access from the customer to obtain these.

This process usually starts with a statement, then an open-ended question:

> **"In order to be sure we are recommending the right
> equipment (service, controls, installation) to best help
> you meet your requirements (objectives), I'd
> appreciate your help in determining key parameters
> of your situation—key physical, numerical and
> interface information. Who should I talk to in order to
> obtain this information?"**

This is a brief explanation of why this information is needed (it's to be sure to get it right for the customer); then the question. An additional benefit of asking the "who" question is that it offers the opportunity to identify and meet decision influencers who might not have been identified earlier.

When talking to the person(s) identified to provide the information, there are specific closed-ended questions to ask about physical and numerical properties. Again, these vary with the equipment and project under discussion. There are also critical open-ended questions:

> **"What are the key parameters for this project?"**
> **"What parameters cover 90 percent of your situations,
> and then what are the extremes?"**

The most effective material handling recommendations often address the 90 percent in one way and handle extremes in a different way. Or, once extremes are identified along with the costs associated with handling them, relative to handling the 90 percent, customers may actually eliminate their extremes or handle them outside the scope of the project under discussion. This often leads a salesperson to a more cost-effective recommendation than competition who reflexively tries to handle all extremes. Alternatively, by asking specifically about extremes, salespeople may help customers understand a situation they didn't know they had—providing a recommendation that others miss.

Other **Establish Parameters** questions include:

If a customer contact provides parameters but seems tentative with the details ("I think..."), the follow-up questions are:

> **"Who can tell us for sure?"**

"How can we verify that?"

Trust, but verify, is critical with parameters. This is where salespeople may need to get out their tape measure, their transits, their counting skills to help the customer determine or verify parameters. Another approach is to ask the customer to confirm with "base" documents such as facility or product drawings, computer-based information, inventory records, or simply by visually confirming information.

Getting the specifications right in material handling is perceived by most customers to be the salesperson's job. Forcing this issue, or personally confirming these parameter details, is the salesperson's responsibility. If the equipment or service or installation or controls don't work, the first recourse is always to the supplier—and ultimately to the salesperson. This is the function customers hold the salesperson most accountable for.

More parameters questions

Additional open-ended questions helpful in the **Establish Parameters** step of the **Objective Based Sales** model include:

> **"What questions did you expect me to ask that I didn't ask?"**

The intent here is to learn something the customer may have held back, even unintentionally, which would affect the specifications and proposal.

> **"What questions did my competitors ask that I didn't ask?"**

This question does three things. It gives you feedback on your competitors and how thorough they have been. It may help you learn something from your competitors. And, if your competitors have been negligent in asking parameters questions, it subtly reinforces this with your customer, again building trust in you for focusing on the customer.

Getting and putting it in writing

In some situations, parameters are so critical that a mistake in this area could cost significant money and time in implementing the project. Ordering a warehouse full of conveyor, forklifts, and pallet rack based on the wrong size load information costs a bundle.

In these situations, particularly when the customer is supplying the information without an easy way for the salesperson to verify it, it can be appropriate to ask the customer to put this information in writing for you.

This will encourage them to do more homework on their end, and make sure it's right.

In the proposal for the project, the key parameter information will be fed back to the customer with a statement like:

> ***"The equipment and services recommended and proposed here are based on the following parameters."***

This statement, with the appropriate operational parameters:

■ Differentiates the proposal from competitors who do not do this.

■ Builds confidence with the customer that the salesperson has made an appropriate recommendation which will help the customer meet their objectives.

■ Provides some protection for the salesperson in the event the information is not correct or changes after order is placed (commitment made!) and implementation made.

■ Confirms the salesperson has performed the function of helping the customer understand what is going on in their own operation.

Before discussing any equipment specifications, **Establish Parameters.**

Focus on the customer and the parameters of their operation, not on the equipment specifications.

Chapter Fourteen

Prioritize and Negotiate Objectives

In many material handling purchases and projects, customers will have more than one objective. Multiple objectives can lead to uncertain or conflicting recommendations. In order to move forward with a purchase in this situation, these objectives must be prioritized.

Similarly, when more than one customer decision influencer is involved, objectives of different decision influencers may conflict. In other words, not everybody agrees as to what's most important. In order for the organization to resolve this and move forward with a good decision, conflicting objectives must often be negotiated among customer decision influencers—and again, be prioritized. Failure of organizations to do this is often a reason decisions are delayed, or in some cases, never made. It's just too difficult to get everyone, as the expression goes, "on the same page."

Alternatively, if a customer makes a decision and purchase without this process, implementation may be poor, not accepted by some, and ultimately reflect poorly on the salesperson involved. In some cases, it results in return of product, or lost customers, or both.

OBJECTIVE BASED SELLING

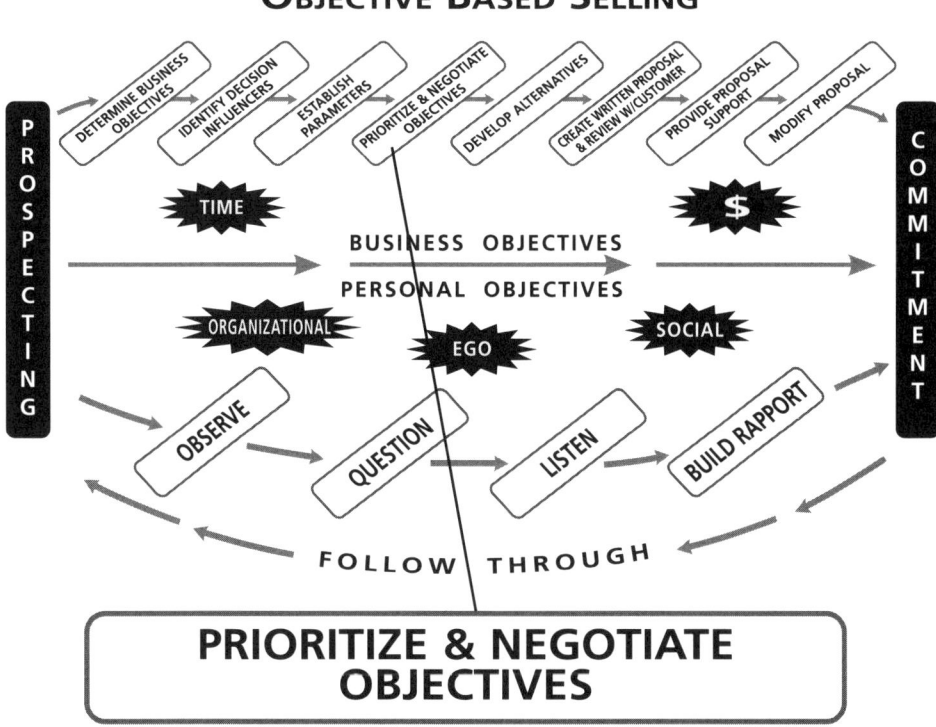

Facilitating the processes of customers' negotiation and prioriti-zation of objectives is a key salesperson function in the material handling, business-to-business sales environment.

In relatively simple situations, with a limited number of priori-ties and decision influencers, this can be accomplished as simply as asking key influencers one of these open-ended questions:

> **"As I understand them, your objectives in this purchase are (__). How would you prioritize these?"**
>
> **"Which of your objectives is most important in this project?"**
>
> **"What's the most important thing about this project/purchase?"**
>
> **"What's the primary thing you want to accomplish with this project?"**
>
> **"What's your biggest concern about this purchase/project?"**

With two or three decision influencers, it is sometimes easy to get them together in a "standing meeting" (perhaps just standing informally together in one office, on the facility floor, or even in a hallway) to ask these questions.

If this is difficult to make happen, each influencer can be asked individually, and the salesperson may be able to build a consensus in this manner.

The result can be tested during proposal review or follow-up, when these objectives are stated back to the customer in writing, following the statement:

"As we understand them, your objectives in this project (in descending order of priority) are:"

Most customers will let the salesperson know if this statement is not accurate.

However, in situations where there are a larger number of decision influencers, and objectives may be numerous, complex, and without apparent general agreement, the consensus building for objective negotiation and prioritization requires more attention, skill, and work.

The very best way to prioritize and negotiate objectives in these situations is to get as many decision influencers as possible together at the same time, and facilitate the process.

One term for this is a "scrum meeting." This term is derived from rugby, where a scrum involves players from both teams meeting on the field, without clear direction. There is a struggle. At some point, a player breaks free from the scrum with the ball, and everyone on the field starts chasing in that direction. The concept is for the salesperson to facilitate a scrum meeting with the result that all customer decision influencers are running in the same direction—hopefully, without the mayhem of a rugby scrum!!!

This is such a critical element of the **Objective Based Selling** process for medium and large size projects, or even smaller projects with longer term consequences, that a separate chapter is devoted to the multiple benefits and purposes of scrum meetings; and how to make them happen as well as facilitate them. That's the next chapter—read on!

Chapter Fifteen

Create and Conduct Scrum Meetings

In the previous chapter, scrum meetings were discussed as a method of facilitating the negotiation and prioritization of objectives by customer decision influencers, so that key decision influencers are all "on the same page" and comfortable in making a purchasing or "project go" decision.

Scrum meetings in that context were defined as a meeting facilitated and conducted by the salesperson, where as many customer decision influencers as possible are together to discuss a significant purchase or project.

While that is a primary purpose (and place in the sales process) of scrum meetings, they are such a powerful sales tool that additional objectives and timing in the **Objective Based Selling** model are appropriate and recommended for scrum meetings. This chapter will deal with the subject in detail, including:

- Why salespeople should want to suggest and make scrum meetings happen

- Why customers will help facilitate, and attend, scrum

meetings (how these meetings benefit customers)

■ When they can be used effectively in the **Objective Based Selling** process

■ How salespeople can make them happen

■ How to facilitate and control these meetings

Scrum meetings are a joint effort by the salesperson and his key contact within the customer's organization. The salesperson is the catalyst, suggests the idea, objectives, agenda, and even some desired attendees. However, the meeting itself must be "pulled together" by a customer contact. They have the credibility, contacts, and inside communication methods to make it happen.

Why scrum meetings: from the salesperson's perspective

Meetings involving multiple customer decision influencers are difficult to put together; involve significant time commitments; and may be risky in the organizational sense to the customer contact who organizes it. In short, when the salesperson's primary contact does this, it's a commitment—to the salesperson who will be there. Getting this done is a "trial close" by the salesperson, an affirmation that the customer trusts him and is willing to invest time. Because scrum meetings require significant time and effort, most customers will not do these with multiple suppliers. Ultimately, if they do a meeting, they will most likely do it only once—and that once will be with the supplier they are "leaning toward."

Since a scrum meeting is a commitment of time, resources, and reputation within the customer organization, it's also a qualifier for the salesperson as to whether or not the project is likely to go forward. If customers schedule and attend a scrum meeting, the project is being viewed seriously within the customer organization, and public commitment of this sort often propels a project forward.

Scrum meetings are also a great way for salespeople to meet additional decision influencers. As a scrum meeting is scheduled and the list of customer personnel attending is made known, key decision influencers will not want to be left out. The list grows. Often, key customer individuals show up unannounced. What an opportunity for a salesperson!

Scrum meetings are effective tools to uncover new information about the project or purchase: additional objectives, key parameters, priorities, even additional projects. During a scrum meeting, sales-people have the opportunity to observe the inner dynamics of the customer's organization. It may also be an opportunity for the sales-person to introduce additional members of his company's organiza-tion to the customer—increasing trust and putting more of a face on his company. This also becomes a chance for additional "eyes and ears" for the salesperson in the project, additional perspectives.

The extended time with customers is a great opportunity to build personal relationships with key customer decision influencers—work-ing on the personal loop of the **Objective Based Selling** model.

Scrum meetings are a key part of the salesperson's function of facilitating decisions, getting everyone on the same page so the proj-ect can move forward. In most scrum meetings the salesperson must avoid overtly "selling." No one will come to a scrum meeting identi-fied as "Sales Pitch," or leave feeling good about the salesperson who turns it into one.

However, the salesperson is of course using the meeting to "sell" his sincere interest in the project, competence, capabilities. The meeting is a basic building block in creating trust that the salesper-son and his company will "get it right" and "get it done."

Okay, why should the customer want to create or attend a scrum meeting?

Instinctively, most customer decision influencers will know that to get general agreement on a project and move forward, a meeting of key decision influencers must occur. However, they often move along in a project without getting it done.

In some circumstances, there is one strong customer advocate for a project or purchase, and they need to create the meeting to "sell" the project to the customer hierarchy. But, they are unsure how to make it happen.

Some customer influencers are looking for opportunities to give their input about objectives, parameters, priorities in a project—and see the scrum meeting as an ideal way to get this done.

Other reasons for customers to create or attend scrum meetings are to spread the risk of a project among other organization personnel; make sure they get the project right; be able to ask suppliers questions in a non-threatening (not one-on-one) environment; get their bosses' attention about a project; or simply to get the project off dead center—to make something happen!

All these are valid reasons for the customer to create and/or attend scrum meetings. These are all benefits to the customer, supplied by the salesperson who makes it happen.

When are scrum meetings appropriate in the Objective Based Selling process?

Scrum meetings can be effective at many points in the sales process. They can be used near the beginning of the process to clarify objectives and key parameters, while meeting decision influencers. As mentioned in the last chapter, they can be used to help customers negotiate and prioritize objectives, with other benefits as mentioned.

What better way to present an Objective Based Sales Proposal to key decision influencers than in a scrum meeting? This is close to decision time, and allows some very pointed questions to be asked of customers (these will be delineated in chapter eighteen, "Review Proposals with Customers").

Scrum meetings, sometimes ad hoc and informal, are pulled together for product demonstrations or more formally, for site visits ("Provide Proposal Support" in the model).

When extensive implementation issues develop during follow-through, a scrum meeting is often the best way to get back on track.

And, in some situations where customers have not yet clearly identified material handling projects, scrum meetings can even be used for prospecting!

The ideal situation for a significant project would be two scrum meetings: one near the beginning of the sales process (Determine Objectives, Establish Parameters, and Identify Decision Influencers) and another for Review Proposals. A salesperson who can make this happen with a project should never lose that order!

In most projects, the salesperson must choose the appropriate timing for the one scrum meeting he will be able to make happen. This, of course, is the art part of selling: making the judgment call that now is the time in this situation for this project.

How to create a scrum meeting

Unless the customer happens to suggest it, the only real way for a salesperson to create a scrum meeting is to suggest it himself.

At whatever the salesperson determines is an appropriate time, the suggestion can take one of several forms, most effectively as an open-ended question (of course!):

> **"What can we do to get key people involved in this project together to review objectives and parameters, so we're sure we get this right?"**

This question suggests the meeting, along with one potential benefit to the customer: being sure "we get it right."

Other scrum meeting catalytic questions might be:

> **"How can we get decision influencers together for the proposal review, so we can be sure our proposal is on target, and that we address everyone's questions?"**

Note, there is no mention of a sales pitch.

> **"What do you think of getting everyone involved in the project together to be sure we have everyone's issues on the table before we prepare a proposal? To be sure our proposal is on target?"**
>
> **"When we make the site visit, it would be helpful to have as many decision influencers there as possible. How can we do that?"**
>
> **"In a project this important, we often encourage customers to pull key decision influencers together to get everyone on the same page. What is your reaction to that idea?"**

and, here, a rare closed-end question:

> **"May I suggest an agenda for such a meeting?"**

In many cases, the customer contact will instinctively understand the importance of such a meeting. In others, the salesperson
must be more assertive and suggestive. Use and explanation of the
term "scrum meeting" can be helpful; it's different and catches the
customer's ear. It is clearly not a sales pitch term.

Offering an agenda for a suggested scrum meeting is another
effective tool for creating a scrum meeting. It sets the stage for the
kind of meeting that will most benefit the salesperson. Specifics of
suggested agendas are discussed in the next section of this chapter.

This is an area where email or fax can be used as effective tools.
Salespeople can create and email (or fax) suggested agendas to customers. In some circumstances this can be done as part of the actual suggestion for a meeting. The customer is offered the suggestion
and encouraged to make it "theirs" by suggesting additions, dele-
tions, rearrangements. It's a working document at this point. They
can also be encouraged to email it to others in their organization,
with suggested time and location. In this way, the customer is building momentum for the meeting, and getting commitments.

The customer contact may take the initiative from here, and
schedule location and time, and invite key influencers. Or, the customer contact may require more assertive recommendations by the
salesperson. Again, this is the "art" of the salesperson, feeling his
way with the customer to make this happen. On balance, usually
the more assertive the salesperson is in this situation, the more likely it is to happen.

A key consideration at some point will be where to hold the meeting. There are basically three possibilities:

- customer location
- salesperson's facility
- neutral facility: hotel, meeting facility, or in some cases, site
  visit facility

The benefits of holding it at the customer location are that it is
more convenient to the customer; it is easier to get more decision
influencers to commit to be there; it offers access to additional information that may be needed as the meeting develops. Sometimes, vis-

its to the appropriate customer facility locations are simply easier.

The disadvantages of a customer location meeting site are that it is easier for customer decision influencers to be late, leave early, or "pop in and out." It can also be easier for a customer to cancel a meeting at the last minute; they haven't all gotten together to travel somewhere, so they just cancel. Also, the meeting room is harder to control at a customer location. They may choose an inappropriate room for what the salesperson had in mind, or without the necessary audio-visual facilities.

Advantages of conducting the meeting at the salesperson's facility or a neutral site are that it is more of a commitment, signifying more interest in the salesperson and his company. The room and meeting facilities can be more controllable. It is easier for the salesperson to prepare and set up. It's also more of an "event," which elevates the importance of the project.

In the end, the location can be suggested by the salesperson, and assertively so; but the customer may dictate the decision of where to hold the scrum meeting.

How to conduct a scrum meeting

Critical elements to conducting an effective scrum meeting are:

- an agenda, prepared ahead of time
- preparation
- control of the meeting from the front of the room
- effective use of open-ended questions to direct conversation
- use of silence to draw out information
- listening and taking notes
- ending meeting with "next steps" or action plan

Scrum meeting agendas

Scrum meeting agendas initially emailed or faxed to customers should all start as some variation of the agenda shown on page 109.

Note: The term "suggested" shows deference to the customer and encourages the customer to modify as appropriate for them. The term "project" should be used on the agenda, even if it has not been introduced as such by the customer. This "upgrades" the purchase and gives it more importance, gives the meeting more importance, and helps the customer compete more effectively for funds and attention in his company.

The term "presentation" is not used, even if the agenda includes a proposal presentation. The better phrase is "proposal review." Again, it is important that the meeting not be seen as a sales pitch. It is also important that the meeting not be a sales pitch.

The first agenda item should always be:

■ Review project objectives

That's the customer focus; that's what gets them interested.

The meeting agenda should always end with "next steps" or "action plan." After all, the primary purpose of the meeting from the salesperson's view is to move the project forward.

Additional agenda items (not shown on sample agenda) may include:

■ Proposal review

■ References

■ Case studies of similar projects

■ Review of salesperson's capabilities

■ Recommendations

Notice that the agenda has a place for attendees for both customer and salesperson's company or team. This is to encourage the customer to add names, and to allow the salesperson to also add attendees. This is a good time for "peer to peer" selling. The salesperson's service manager talks to the customer person in charge of service; finance person to finance person; etc. As the customer's list of attendees expands, the salesperson may want to invite more from his sales team.

Suggested Meeting Agenda
Warehouse Modernization Project
XYZ Corporation

☐   Review of project objectives

☐   Key parameters

☐   Priorities

☐   Concerns

☐   Time frame

☐   Financial considerations

☐   Questions

☐   Next steps

Attending for ABC Material Handling Sales Company:

Names and titles

Attending for XYZ Corporations:

Names and titles

Suggested date, time, location:

Preparation

Due to the importance of the meeting, preparation by the salesperson and his team attending is critical.

After selecting the sales team members to attend, a pre-meeting should be held. One person, probably the salesperson, should be designated as in charge. This is critical. There must be clear direction by the meeting leader and chair, and others on the sales team should actively participate only when directed or invited.

PowerPoint or other audio-visual tools should be prepared and given a trial run. Get it together before the meeting!

Conducting the meeting itself

Arrive early, set up, do a test run-through of any audio-visuals.

Water is okay, but try to avoid food or meals, as this only complicates the meeting. If food must be served, use it as a discussion time after the meeting—not before or during the main agenda part of the meeting.

Direct team members as to where to sit. It is best to space your team around the room, so it doesn't become customer on one side of a table or room, salesperson's people on the other. This suggests "us" versus "them" mentality; what the salesperson is striving for is a "we" environment.

The salesperson's team member in charge should facilitate introductions as customer people enter the room, being clear on names and asking customers two key open-ended questions as time and arrangements permit:

"What is your area of responsibility?"

Remember, never ask title; get that from their card.

"What is your relationship to this project?"

Be prepared to start the meeting on time, but start only when the primary customer contact says to go ahead. If necessary, encourage a short delay in order to get as many decision influencers in the room at the start as possible.

When additional individuals enter during the meeting, the meeting leader should stop, welcome them, introduce himself, ask who the new person is (if not known by all present), and ask the two questions again:

"What is your area of responsibility?"
"What is your relationship to this project?"

It's also appropriate to succinctly summarize what's taken place in the meeting so far. "Catch up" the new person with the meeting.

One of the most important techniques in conducting—and controlling—a scrum meeting is to never, *never* hand out materials before the meeting starts. This is distracting; it may encourage some to leave *(Well, I've got the material and I'm busy)* before the meeting even starts; and it will be almost impossible to get customers' complete attention when the meeting starts *(They gave me this stuff, so I better read it)*.

This is important in all scrum meetings. It is absolutely critical in a scrum meeting designed for proposal review. If customers are given a proposal before a meeting starts, they will look immediately for the price. When they see the price, selling stops.

To control a scrum meeting from the front of the room requires use of PowerPoint (or overhead projector) and open-ended questions.

After introductions and a review of the meeting objectives, the meeting leader should review the meeting agenda (again, on PowerPoint or overhead first; then it's okay to hand out paper copies).

Before proceeding, the leader should ask the following questions:

"What is not shown on the agenda that someone here
would like to be sure we cover?"

Whatever the answer, it should be written down and covered before the meeting is over.

"What is the most important thing about this project?"

Even before jumping into the agenda, the leader should encourage attendees to say what's the most important thing on their mind.

The meeting leader should direct most open-ended questions, particularly at the beginning of the meeting, to the group at large— not to any specific individual. This allows the opportunity for people to get out any really burning ideas, thoughts, or questions. It also allows the salesperson to observe the dynamics of the customer's decision influencers.

After facilitating this initial discussion, the leader should redirect attention to the agenda.

Agenda items should be introduced with open-ended questions, or introductory statements followed by open-ended questions.

> **"As we understand them, XYZ's objectives in the Warehouse Upgrade Project are (insert bullets with PowerPoint). How on target are we?"**

"What have we missed?"

"Which are the most important objectives?"

And so on for other agenda items.

"What are the key parameters?

"What parameters have we missed?"

"What are key project time frames?"

One rule of thumb for scrum meetings: Customers should talk about 80 percent of the time, salesperson's team about 20 percent of the time.

If one customer person begins to dominate the discussion and others become non-participants, the leader should redirect conversation with a new question directed to the group, or to someone specific. One such question might be:

"What are others' reactions to this issue?"

As customers answer the meeting leader's questions, the sales team should follow up by following the trail of the answers with... more questions!!!

Customer discussions among themselves should be encouraged and seldom interrupted.

Handouts should be delayed to the end of the meeting, or to points in the discussion where they are referenced.

Summarizing, as it is time to end the meeting:

"Let me summarize where I understand we are." Here the leader can quickly recap key objectives, parameters, issues learned in the meeting.

Questions to end the meeting are:

"What's the next step in the project?"

"What would you like us to do next?"

"What's the time line from here?"

"What are the steps in our action plan?"

Effective facilitation of customer scrum meetings can only come with practice. These guidelines are the starting point for one of the most effective sales tools available for significant material handling projects. One reason the scrum meeting is so effective is it performs valuable functions for the customer.

More specifics about scrum meetings for proposal reviews with customers will be given in chapter eighteen, "Review Proposals with Customers."

Chapter Sixteen

Develop Alternatives

Customers naturally want to compare alternatives: alternative brands, models, methods, dealers, warranties, layouts, financing, timing, and, of course, pricing. This is done to give perspective to their decision; to be sure they have all the relevant ideas; to satisfy their boss; and, to use an old sales cliché, to "keep the salesperson honest" (in other words, to beat down the price!). In larger organizations, controllers, finance managers, purchasing agents are all paid to ask the questions "Have we considered (_____)?" or "Have we compared (_____)?"

In material handling operations, purchases, products, projects, and methods, there is always an alternative way to do things. It might be a different model, layout, financing, used versus new... the industry is built on comparing and selecting the best alternatives.

This leads the material handling salesperson to develop alternatives so the customer is considering and potentially choosing between two of his alternatives rather than a competitor's alternative. In situations where relationships are strong and time is short, providing and proposing alternatives may in fact lead the customer to avoid talking to competitors at all.

OBJECTIVE BASED SELLING

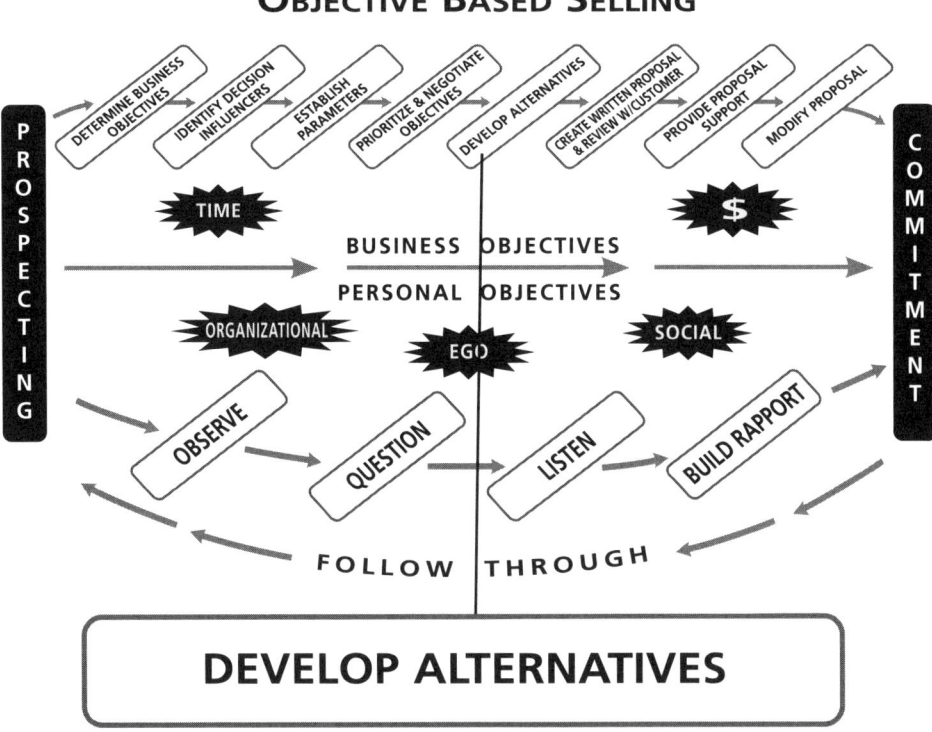

DEVELOP ALTERNATIVES

Think alternatives from the initial contact

With this concept in mind, material handling salespeople should be thinking about alternatives from their initial contact with a project or purchasing situation. The more thorough the process of identifying objectives and parameters, the more creative the salesperson can be in helping the customer compare the relative benefits of alternative methods at alternative price points. In material handling, alternatives offered may include:

- Lesser models with fewer performance features; or more sophisticated models with more performance features (or, differing brands with the same distinctions mentioned here)
- Options to perform added functionality
- Alternative financing: leases, lease with maintenance, shorter or longer lease terms, leases with buyouts, "walk away" leases, full payout financing, leases with different residuals and end of term alternatives; the possibilities with financing are endless

- New versus used equipment
- Trade in old equipment, versus keep it for back-up, or use it in another part of the customer's operation
- Buy part of a requirement or quantity now, part later
- Varying warranty lengths, coverage (again, the alternatives here have become very creative and varied with material handling equipment)
- Freight included, or customer provides the freight
- Alternative layouts
- Capacities reduced, to handle 90 percent of loads or situations; handle the other 10 percent a different way
- Manual versus powered models
- Installation provided versus customer assumes installation responsibility
- Buying more now to get a price break
- Buying less now to meet budget restrictions
- Totally different way of looking at project to achieve customer objectives. Example: conveyor vs. forklift; manual controls vs. automated software; electric vs. propane powered forklifts; hydraulic vs. mechanical dock levelers; modular drawers vs. shelving; bulk rack vs. shelving; wood shelves vs. steel shelves; and on and on

These ideas can be explored with customers before committing the resources to offer them in detailed proposals. Open-ended questions that help with this are:

"What if we (_____)?

"What is your company's attitude toward (_____)?"

"What financing alternatives have you considered?"

"What features would be dispensable if we lowered the investment significantly?"

"What if we lowered the investment significantly by handling 90 percent of your loads this way, and the other 10 percent on an exception basis?"

"What alternatives would you be willing to consider if the proposals all come in over budget?"

"What capabilities would you consider compromising in order to significantly reduce the investment?"

"What capabilities would you like to improve over the
basics? Why?"

"If we could do it, what would you be willing to pay more
for?"

"What one thing, if we added it to this project, would
dramatically improve the way you do business?
Lower your costs? Improve customer service?"

And, while not strictly speaking an open-ended question, the following phrase can be an effective tool in checking out a "trial balloon":

"Think through this with me. If we (_____)"

When offering alternatives:

Alternatives should always be at different price points. This
gives the customer the empowerment of spending more to get more;
or spending less to indicate prudence and show their company they
are saving money.

Alternatives proposed should always be accompanied by clear
explanations of how the alternatives differ in performance, differ in
helping the customer meet their objectives, or deal with operational
parameters.

"With this model, for this price, we meet all your basic objectives
and operational parameters. For an additional investment of 12 percent, we can help you deal with the expected changes next year, and
provide these additional safety features. Which is more important to
you: the lower investment now, or the additional safety and future
flexibility?"

The customer is choosing between two alternatives—both yours!
Whatever the choice, he can tell his boss that he considered two
alternatives; this is what he recommends, and why. No real need for
him to talk to a competitor, is there? And, no immediate need for the
salesperson to just drop price and reduce margins. Instead, let's consider a lower-priced alternative!

How many alternatives?

Every significant material handling proposal should offer at least
two alternatives, even if one is only an option to be added or deleted.

The more significant the project, the more complex the situation, the more decision influencers, objectives and parameters to deal with—the more the material handling salesperson should consider offering additional alternatives. One reason for this is to give the customer more realistic choices. Another reason is to anticipate alternatives that might be offered by competitors. A problem with this, of course, is the amount of work which may be involved in pulling together meaningful pricing and proposals for varying alternatives. The salesperson must evaluate whether there is payoff in this or not. That is where the trial balloon questions mentioned earlier in this chapter come into play. Will the customer really consider these alternatives? Will the differences be meaningful?

A second problem which can develop with more than two alternatives is customer (and salesperson!) confusion. There simply may be too much to consider, particularly if not clearly presented.

One technique for clearly presenting more than two alternatives is a matrix or spreadsheet presentation. One axis has the different models; the other has the different implications for the customer, along with investment differences. Another possibility for presentation is a simple one or two paragraph summarization of alternatives and associated implication differences for the customer.

However, if the customer is not really open to considering all the alternatives offered; if the explanation of what is different for the customer about each alternative is not clear; if the effort to put the additional alternatives in proposal form does not appear to be worth it—better to only offer two alternatives.

In the end, which alternatives to offer, and how to present them to the customer is a judgment call of the salesperson. That's the "art" part of selling!

But remember: Someone at the customer's organization will probably be asking, "Have we considered (_____)?" "How can we lower our investment?" "What if (_____)?"

If the salesperson doesn't offer the customer alternative methods, models, payment structures and pricing... competitors will!!!

Chapter Seventeen

Create Written Proposal

Almost all significant sales of material handling equipment and services are ultimately actualized on the basis of a written proposal from the salesperson and selling company, to the buying organization. This is done for the benefit of both the customer and the selling organization.

Customers require written proposals for a number of reasons, including:

- Clarification of relatively complex details of equipment, services, warranty, pricing, terms, delivery time... the entire sales offering
- Sharing the information with appropriate decision influencers within their organization
- Comparing competing proposals
- Maintaining information over time
- Ultimately, legal and commercial reasons

"Written proposals" can refer to actual paper documents, or digital versions, or both.

Selling organizations and salespeople have traditionally offered written proposals in order to:

OBJECTIVE BASED SELLING

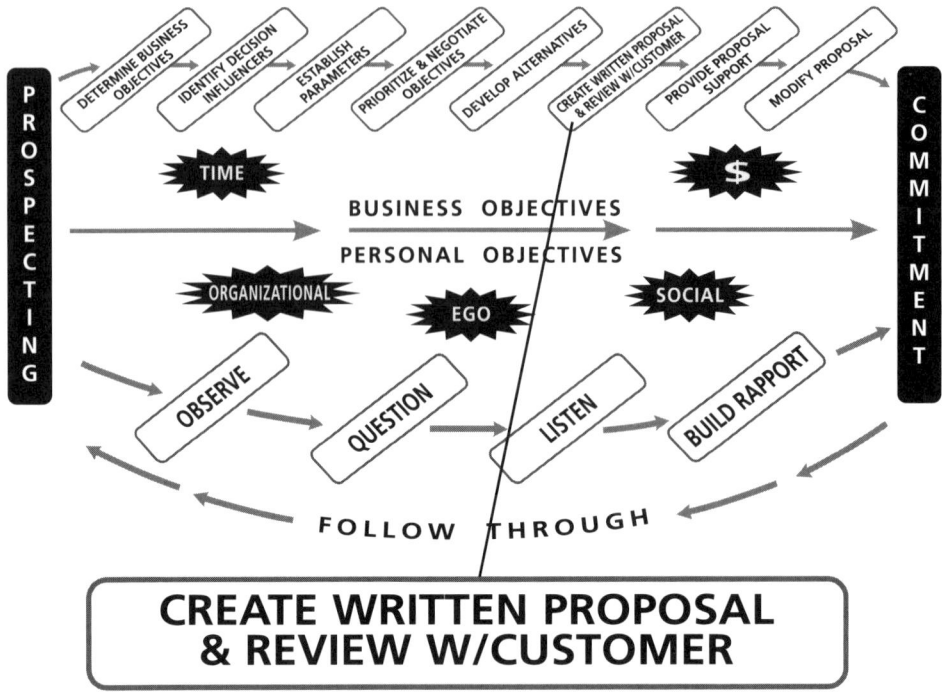

- Meet customer requests and requirements for them
- Clarify relatively complex details of their offer
- Get it on paper
- Avoid misunderstandings
- Comply with legal and commercial reasons

It is the way most business is done.

The format of most traditional material handling sales proposals has been, and remains, a detailed statement of equipment specifications; price; perhaps some options priced separately; delivery information (usually vague or inaccurate); a brochure; payment terms and conditions; a quick sales pitch, usually based on the concept that "mine has more features so you better buy from me."

The emphasis in these documents is almost always equipment and price. That is why these have traditionally been called "quotes" (or worse, in some situations, called "bids"). They are quoting documents, not selling documents.

The **Objective Based Selling** model, however, emphasizes additional purposes and functions of written proposals, from the salesperson's perspective, and also to help the customer. Recognizing these additional, critical functions of sales proposals leads to a sales proposal format which is different from traditional formats—and more effective for the salesperson and for the customer organization. From the salesperson's perspective, sales proposals in the **Objective Based Selling** model have the following functions, in addition to those listed above:

■ **Sell when the salesperson can't be there**

As emphasized throughout the **Objective Based Selling** model, decisions in the material handling sales environment are most often made when the salesperson is not in the room. What speaks for the salesperson then? Two things: the written sales proposal, and the primary customer contact—who is usually not a salesperson and who may have limited sales skills. In **Objective Based Selling**, the proposal is not a quoting document; it is a document that sells when the salesperson can't be there.

■ **Convey information to hidden decision influencers**

In many situations, no matter how hard he tries, the material handling salesperson cannot meet all the decision influencers. They may be out of town; shield themselves from salespeople; be involved in the project at the final stages; be new to the company or area of responsibility involved with the project. There are many reasons. Again, when the salesperson cannot meet a decision influencer, what sells for them? The written proposal and his primary contact.

■ **Provide a sales tool for the salesperson's primary customer contact**

Even when salespeople obtain the endorsement and recommendation of their primary customer contact(s), they must still convince other, often hidden, decision influencers. Yet, most customer operational contacts with whom material handling salespeople are dealing are not salespeople themselves and may have poor selling skills. They need a sales tool that almost speaks for itself. That is another function of the written proposal.

■ **Compete for funds**

The salesperson's offering may be the one selected, but the order may still be at risk because there is almost always competition for capital and service funds within customer organizations. Someone in the customer's organization is paid to ask: "Can we skip this project altogether?" "Do we have to do this project this year?" "Why should we do this project at all?" "Will this project pay for itself?" "What does this purchase really do for us?" The salesperson's proposal should help answer these questions before they are asked.

■ **Compare alternatives—both yours!**

As mentioned earlier, there are customer decision influencers paid to ask "How can we do this for less money?" "What else have we considered?" "Have we considered other alternatives?" "Have we considered (_____)?" The effective sales proposal helps answer these questions by offering at least two alternatives, with clear differing benefits, at two price points.

■ **Provide reference and structure for the personal proposal review, (when getting the opportunity)**

Many salespeople are strong with interpersonal skills, but not good presenters. A well-constructed written sales proposal can easily serve as an organizational outline for a personal proposal review or scrum meeting presentation. For a scrum meeting presentation, proposals can be copied for overhead projector use or summarized and converted to PowerPoint for electronic projection.

■ **Explain the whole proposal—your entire way of doing business—not just product and price**

The path to competing for business at higher gross margins is to offer a more complete way of doing business: the right product, the right package of services, warranty, follow-through, and back-up. This must be explained in this sales document. The customer should be made aware there's more to the salesperson's offer than product and price!

■ **Present the proposal in a way that stands out from the competition, that effectively competes, that sells**

Simple presentations of product specifications, price, and delivery do not compete. They quote. Low quote gets the deal. That results in either lost deals or lost gross margin.

■ Motivate action!

Salespeople who "wait for the customer to do something" are not earning commissions. Many material handling salespeople spend time and company resources working on customer requests or projects that never happen. "We decided not to do that project this year." This is the same net effect for the salesperson as a lost order. The written sales proposal should be seen as an opportunity to be a catalyst for customer action on the project or purchase. It should help answer the implied customer question: "Why should we do this project—and do it now?"

These are critical, non-traditional functions of written sales proposals for salespeople in the **Objective Based Selling** model. In our customer-focused model, what are the functions of written sales proposals *for customers?* Well-constructed sales proposals:

- ■ **Help primary customer contacts and project champions compete for funds in their own company**

- ■ **Clarify which sales proposal will most completely meet their objectives**

- ■ **Provide support for a supplier recommendation; provide trust that the selling organization can in fact get the job done**

- ■ **Pull together their own objectives in a prioritized manner, with key parameters outlined in a manner that may not have been done within the seller's organization**

In other words, effective written sales proposals in the **Objective Based Selling** model perform key functions for both the salesperson's and the customer's organizations. Proposals which do this are called **Objective Based Selling** proposals. They are, in fact, customer-focused proposals.

Differences: traditional versus Objective Based Selling proposals

The differences in format between traditional sales proposals and **Objective Based Selling** proposals are:

- **Objective Based Selling** proposals start with the customer, instead of the equipment (or service) being offered

- **Objective Based Selling** proposals focus on customer objectives and parameters—the job to be done—rather than simply the equipment and price

- **Objective Based Selling** proposals are constructed as selling documents—both to the customer and within the customer's organization—rather than quoting documents

- **Objective Based Selling** proposals help explain how making the purchase (or completing the project) for the equipment or services being proposed will benefit the customer, and how the offering in the proposal is the right one to help the customer meet their objectives

Objective Based Selling proposals tell a story of why this project is important to the customer; what the customer gains from it; under what set of circumstances (parameters); with what alternatives to consider; at what price points.

Additionally, **Objective Based Selling** proposals should identify benefits of buying from the company offering the proposal; where that company has successfully done similar projects; any time, job, compliance, parameter, or qualifying issues; and the scope of the entire sales proposal—not just product and price.

Key elements of Objective Based Selling proposals

Objective Based Selling proposals of significance should have the following key elements:

■ Frame

This should be both physical and metaphorical. Its purpose is to say to the customer, "Here comes a proposal of significance from a professional company. Pay attention."

Where the proposal is an actual paper document, the physical frame can be a proposal folder, binder, or similar identifying device. It helps upgrade the proposal from a piece of paper to one of more significance.

Where the proposal is a digital document, a cover sheet with sales company and customer logos can be created. This should include project name, date, and other information that adds a professional appearance.

The metaphorical frame is a one-page letter of transmittal. This should accompany all but the simplest sales proposals. A suggested basic template is given in Appendix 2. Important features are confirming or giving the project a name; thanking the customer; inclusion of key decision influencers' names; table of contents of the remainder of the proposal; basic recommendation statement; summary that the proposal is based on a lot of work the salesperson and his company have done. Remember: This document is talking to all decision influencers who will see it, not just the addressee.

Adding a P.S. with appropriate, relevant information is good. P.S. statements are almost always read, sometimes before the body of the letter!

There should be bullet points and lots of white space. If it is a paper document, there should be an actual signature. (There is something about actual signatures that helps make a personal connection.) If the table of contents is more extensive, it can be referred to in the letter of transmittal, and then made a separate page.

The letter of transmittal should never be more than one page, and should have the offerer's (usually salesperson's) name at the bottom, even if the letter is electronic (signatures can often be scanned into electronic documents). This helps personalize the offer. If an engineer, manager or other person from the salesperson's company was heavily involved, a second name (and signature if practical) should also appear on the letter of transmittal. This letter sets the professional tone and standard for the remainder of the proposal.

■ Summarized statement of customer objectives

This is one of the most distinguishing characteristics of **Objective Based Selling** proposals. Most material handling sales proposals don't bother restating what the customer is trying to accomplish with the purchase or project. It's a big omission. It's the reason to do this project at all. These customer objectives should be stated in summarized bullet form, following a statement such as:

"As we understand them, XYZ's objectives in the Warehouse Upgrade Project are:"

Following a listing of the objectives, the simple statement should be made:

"These are listed in our understanding of their order of priority."

What better place to begin a sales proposal than with the objectives of the customer?

■ Listing of key customer parameters

Again, these should be in bullet form, following a statement like:

"Key parameters of XYZ's operations for the Warehouse Upgrade Project include:"

These are the circumstances understood to be important in achieving the customer objectives. These should be as specific, and numeric, as possible. They also serve as some protection for the salesperson if the parameters later change, causing problems with the equipment recommended, or if the salesperson has been given inaccurate information.

Listing customer prioritized objectives and key parameters are a major service for the customer in helping them sell the project internally—and a major differentiator of a sales proposal. Why? Because most sales proposals don't offer this information.

■ Alternatives offered

Without pricing, this section of the proposal should explain alternatives considered for recommendations, with clear explanation of how these differing alternatives vary in meeting customer objectives. Again, the primary concern is meeting customer objectives within the given parameters, not detailed equipment specifications.

■ **Specific recommendation**

If the salesperson believes one of the alternates offered is most appropriate for the customer, that should be stated clearly, with reasons why. Reasons should relate to customer objectives and parameters.

■ **Equipment specification sheets; drawings; other details of the offering**

These should be highlighted in some manner, including now any distinguishing characteristics of offerer's equipment or services—but again, highlighting how these distinguishing characteristics relate to customer objectives and parameters.

■ **Benefits (to the customer) of acting on your proposal**

Here's where the salesperson gets specific on how his proposal best helps the customer meet their objectives within the specific parameters of their situation.

■ **Bill of material, with quantity, specifications, investment and commercial terms**

Okay, it's time to get specific, including the price. The price should be listed as an investment—implying the customer gets something back for it, it's not just an expense.

■ **Delivery information**

Unless including a reverse time line (explained in chapter twenty-two, "Obtain Commitment"), it is recommended no information be included on delivery. Reasons: Chances are it will change before the proposal is acted on; it's almost always inaccurate; except in rare cases, it will not positively influence the selection of the supplier at this point, and can work against the company proposing—e.g. if current delivery information is way off customer objectives, they may disregard the proposal for this reason, when this very information can be impacted later by sales or supplier actions. Also, by not including delivery information, the salesperson opens up an avenue for future conversations, when the customer asks about it.

■ Proposal support

Never end a proposal with the price. Once the bill of material and investment proposal have been given, it's time to give "proof statements" about how your company can be trusted to do this project and help the customer meet their objectives within the parameters of this situation. Proposal support should always include a list of similar jobs successfully completed, or referrals. Always. If permission has been obtained in advance, you can include contact names and information. If this is not available, references or past jobs can be described generically. But they should always be included.

Other proposal support can include insurance certificates (if seller's personnel will be on customer job site); seller's organization charts; pictures of previous jobs; lists of suppliers; lists of awards earned by selling company; the list is almost endless. The key: The list should be tailored to the customer and to their specific situation. Only proposal support relevant to this customer's current project, objectives, and way of doing business should be included.

Summarizing
Objective Based Selling
proposals' key elements:

■ **Frame**—physical and metaphorical; folder, letter of transmittal

■ **Statement of customer objectives**—prioritized

■ **Parameters of customer situation / application**—specific, numerical if available, verified

■ **Alternatives**—with specifics as to how differences impact customer objectives, parameters

■ **Recommendation**—Based on all this, what does the salesperson recommend for these objectives in this situation?

■ **Literature / drawings**—highlighted as relates to this customer's specifics in this project

■ **Benefits**—of these recommendations to this customer, related to these objectives, in this situation

■ **Bill of material, investment proposal**—details about quantity, investment, specifications, commercial terms

■ **Proposal support**—proof statements, building trust with customer that offerer can meet customer objectives with these recommendations; should always, always include references

Additions and variations

Some proposals are so important, significant, large, or complex that additions above the key proposal elements are justified and needed.

These can include:

■ Statement of how customer's job is being done now, or what is currently being used. This is helpful where dramatic physical or operational changes are proposed or major brand changes recommended. This helps set the stage for decision influencers not totally familiar with current procedures or brands. It also allows the salesperson to point out important deficiencies with current methods or equipment. When used, this should be relatively succinct (bullet form is always helpful) and positioned in the proposal before the customer objectives.

■ List of key members of sales follow-through team. This is proposal support which helps build trust where subcontractors are to be used in implementation, or where sales team includes experienced project or service managers who are sales points for the selling company.

■ Time line for implementation. Reverse time lines are best (discussed in detail in chapter twenty-two, "Obtain Commitment")

■ Before and after drawings

■ Proposal summary. This should be used for complex proposals with many different types of equipment proposed and priced separately, or complex proposals with several phases. The overall picture can get lost, and should be summarized, along with summary investment numbers.

■ Alternative spreadsheet. Where more than two alternatives are proposed, or several, complex options or alternatives, a spreadsheet can be a helpful form of summary.

■ Cautions. These can be included where customer has mentioned implementing solutions that may have operational, time frame, or even safety risks. Or where customer activities not included in the proposal are required for successful implementation (training, for example); or in situations where there are unknowns prior to starting the project. Examples here might be condition or adaptability of current customer equipment being reused as part of the project; building permit authority actions; facility issues such as floor condition; importance of not exceeding key parameters on which the proposal is based.

■ Executive summary. Sometimes added to particularly complex or long proposals. Many top decision influencers will only read this. Powers of summarization are important here.

How long should the proposal be?

Legend has it that Abraham Lincoln was once asked, tongue in cheek, "Just how long should one's legs be?" He is reputed to have answered, "Long enough to reach the ground."

How long should **Objective Based Selling** proposals be? Long enough to get the order. This, of course, is a judgment call based on the specific situation, including how large the job or project is, or how important it is. The list of elements of **Objective Based Selling** proposals is not meant as a length measure, but as an effectiveness measure. All of the above can be done in three pages; most can be done on a one-page proposal.

One problem with one-page proposals is they eliminate the letter of transmittal and focus immediately on the price, which is showing. However, there are situations for which this is appropriate.

By pre-structuring word processed templates, salespeople can quickly choose and begin to structure the appropriate length proposal, keeping all or most of the key elements.

The important thing is not the length of the proposal, but its effectiveness.

Does it sell, or just quote? Is it a bid, or a proposal?

Including the key elements of **Objective Based Selling** proposals assures it will be a distinctive, selling document.

Summary

Customer-focused, **Objective Based Selling** proposals can be key differentiators in business-to-business selling situations. They speak for the salesperson to hidden decision influencers, and in meetings salespeople cannot attend.

Proposals should be constructed using the key elements listed in this chapter as a guide, with ultimate proposal criteria of:

■ Customer focus, not product focus

■ Selling document, not quoting document

■ Facilitates customer decisions

■ Catalyst for customer action

Proposal templates

Appendix 2 of this book has three suggested **Objective Based Selling** proposal templates: a Long Form for larger, complex, more significant projects; the Proposal Sandwich, for significant but simpler projects or purchases; and the Simple, One Page Form for "everyday" use. These are meant to be used as guides and starting points; again, to be customized by individual salespeople and their companies.

These do not include the commercial terms and conditions items, which are usually pre-printed as a proposal insert, or in many cases, on the back of a proposal form which is used for bill of material or investment proposal.

Proposal is ready; now what?

The next chapter discusses why it's so important for salespeople to get the opportunity to review in person their proposal with the customer, with appropriate customer decision influencers; how to do this effectively; and what to do if denied this opportunity.

Chapter Eighteen

Review Proposal with Customer

The most effective way to present or review a sales proposal with a customer is in person. A personal review allows salespeople to gauge the customer's reaction to the proposal; ask questions; answer questions; obtain trial close agreements and in certain circumstances, close the order. Personally reviewing a proposal with a customer is prime selling time.

Yet, customers routinely tell salespeople to:

- "Just give me the price over the phone."
- "Fax the quote over."
- "Mail it."
- "Leave it in the lobby, I'm too busy right now to meet with you."
- "Email it."
- "Just leave it with me; I'll call if we have questions."

This happens for several reasons, including because customers have very crowded schedules; and because they believe salespeople will just be giving them a sales pitch they don't want to hear or trying to pressure them into an order with inappropriate, traditional

OBJECTIVE BASED SELLING

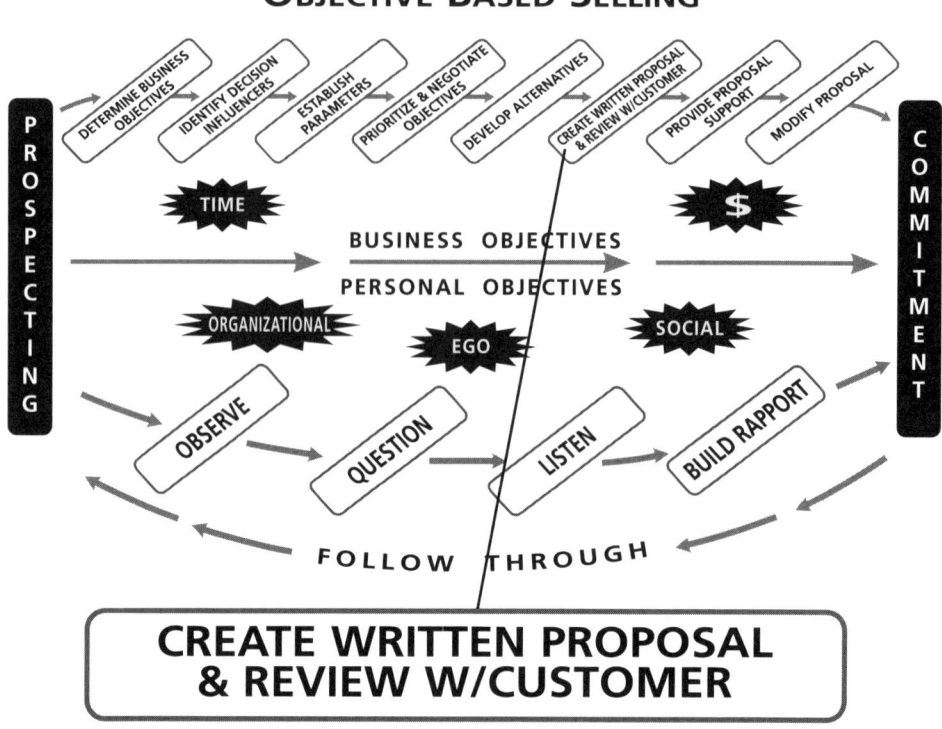

closing techniques. And, of course, customers are right about that in many situations with salespeople who don't have **Objective Based Selling** skills!

Problems with phone, email, fax, other forms of proposal delivery

Other than eliminating the opportunity for personal interface, communication, and selling opportunities, there are further problems with forms of proposal delivery other than in person. Here is a summary of the most obvious.

Giving the price over the phone: This leaves out many details, is probably just the customer's way of using your price to quickly compare numbers with competitors, and allows only the price to be transmitted to other decision influencers. This is a prelude for later misunderstandings and mistakes if the salesperson does get the order.

Fax: Often faxed to a wrong number; may be seen by unauthorized personnel at customer location; not easy to fax color; at mercy of customer faxing and delivery mechanisms for quality of paper; pages often put in different order than intended, spoiling the impression meant to be given; fax machine may be out of paper, causing delay or non-delivery; faxes often go to wrong numbers; without calling to confirm, can't really be sure the appropriate person gets it; can't effectively fax a brochure or odd-sized drawings.

Email: Customer spam control may stop it; may be deleted inadvertently; may not be recognized by customer; if graphics included, customer may not have capacity to receive; can't easily transmit outsize drawings; color may not transmit well; customer may not have appropriate software to open all documents, particularly drawings; brochures can be difficult to email. One advantage of email is it is easy for customer to retransmit to other decision influencers; email transmission following personal review can help accomplish this purpose.

What's a salesperson to do?

Despite customer tendencies to the contrary, salespeople should fight for the opportunity to review sales proposals in person with customers. They should do this because it's the effective way to sell; because it's ultimately in the customer's own best interests; and because they have earned the right to do this, by working professionally through the **Objective Based Selling** process.

This chapter will discuss how to create these personal proposal review and presentation opportunities; how to effectively do the proposal review—whether with a single individual, small group, or in a scrum meeting; and what to do if, in spite of best efforts, the salesperson just can't get into a personal proposal review situation.

How to make personal proposal review meetings happen

The primary way to obtain personal review time is to expect it and ask. Groundwork can be laid for this during earlier meetings, with statements like:

"When we review the proposal together…"

Requesting a proposal review meeting is similar to requesting a scrum meeting, and in appropriate instances, can be converted to a scrum meeting for proposal review. One tip: Use the word review instead of presentation. It's a less threatening word and defuses the sales pitch issue.

"Who should be there when we review the proposal?"

"How can we get appropriate decision influencers together to review the proposal?"

Pointing out the benefits to the customer of personal review, or scrum meeting for review, can improve the chances:

"This will give you the opportunity to assure the proposal is on target, and to ask questions."

"A scrum review meeting can help get everyone on the same page, and perhaps move the project along for you."

Even a little humor can help.

"I'm not doing my job if I don't review this proposal with you."

"My company fires salespeople who don't review proposals personally with customers."

As with other scrum meetings, an emailed or faxed "tentative agenda" can help create the proposal review meeting. The proposed agenda should focus on customer objectives, parameters, issues, concerns, time frame... indicating it's not a sales pitch.

Whether meeting one-on-one to review a proposal with a customer, or in a scrum meeting, basic procedures are the same for an effective review.

- Control the document; don't just hand it to customers.

- Before starting the review, ask several specific, appropriate open-ended questions. (These will be covered in the next section.) Goal: Twenty minutes of questions and discussion before even starting review of the document.

- Use the document as the notes for your review presentation.

- Seek agreement at each section of the proposal—these are trial close agreements.

- If review of earlier sections of the proposal reveal the specific proposal in hand is indeed inappropriate (due to unforeseen changes, new information, competitive information, or customer statements), stop the review and ask permission to take the proposal with you, modify it, and bring back a more appropriate proposal. You don't have to leave it just because you brought it!

■ Ask open-ended questions as you go, and particularly at the end, to get customer reaction. These questions will also be covered in the next chapter section.

■ Even if the situation appears not to be right for a customer decision at that moment, try at least one close anyway. If it's clear there won't be a decision today, this can be tempered with some humor.

Open-ended questions for proposal review meetings

Proposal review meetings are among the most critical times for asking open-ended questions. The customer is focused on the project, the salesperson, and his proposal. He (or they) are closing in on making a decision on whether to go forward with the project, and if so, which supplier proposal to choose. The salesperson benefits by encouraging customers to talk about the project, the decision-making process, the time frame, and their reaction to the proposal.

Unfortunately, most salespeople believe this is their time to talk. And, of course, some talking, verbal review, or outright presentation by the salesperson is appropriate. However, for maximum effectiveness, it's also time to ask questions and listen to the customer. As with most other parts of the sales process, *questioning and listening* is still more important than telling and talking.

The following open-ended questions all may be appropriate for proposal review meetings. Their purpose is implicit in the questions. The salesperson should ask as many of these as seems appropriate (one goal: twenty minutes of questions and answers) before beginning proposal review; ask other questions as the review progresses; and critical questions at the end. In some situations it may be appropriate to repeat certain questions at the end of the review, to determine if answers are still the same. Many of these questions will have been asked earlier in the sales process; again, it's okay to repeat them to determine what may have changed or to reinforce the answers in the customer's mind. The salesperson should follow the trail of the answers, listening to answers and asking appropriate follow-up questions. Begin or continue proposal document review as customer starts to get restless, or as it simply seems appropriate. Let the customer signal when the meeting should end.

Open-ended questions for proposal review meetings, listed in sequence appropriate for many such meetings, include:

"What's changed since we last talked?"

"What's the most important thing about this project?"

"What is your biggest concern about this project?"

"What would you like me to focus on in this proposal review?"

"What criteria will determine whether this project moves forward?"

"What's your decision-making process following this meeting?"

"Who, besides yourself, will be involved in the decision process? What will your role be?"

"Who is the project champion?"

"Who is against the project?"

"What will your criteria be to select a supplier?" The answer to this question is often "price." When that is the response, one follow-up question might be:

"If all the prices were the same, what would your criteria be?"

"What will happen to this document after this meeting?"

"What is your reaction to our proposal?"

"What surprises you about this proposal?"

"How does this proposal meet your objectives?"

If the salesperson is aware competitive proposals are being considered, questions might be:

"What did you see in competitive proposals that you liked? That surprised you?"

"What should we be doing to follow up with you

following this meeting?"

"What is your time frame to move this project forward?"

When prepared with an **Objective Based Selling** proposal, the salesperson should maintain focus on the questions and on listening to customer responses.

Reviewing the proposal document with the customer

Of course, in addition to asking questions, the salesperson does need to review the proposal document with the customer.

In one-on-one meetings, the salesperson should assertively arrange seating so he is sitting next to the customer, or between them if there are two people there.

The document should not be handed to the customer. It should be controlled throughout the review process by the salesperson, and reviewed page by page. This avoids the customer jumping directly to the price.

Selling stops when the customer gets to the price. The customer is too busy doing math or comparing with budgets or competition. In fact, when the salesperson gets to the price page, hopefully well into the meeting, a good tactic is to simply be silent while the customer thinks about it a few seconds.

Another advantage of **Objective Based Selling** proposals is the meeting doesn't end on price. It hesitates there, and then moves to the references and other forms of proposal support.

In proposal review meetings with more than two customer persons—in essence a scrum meeting—the meeting should be controlled by reviewing the proposal from the front of a room with an overhead projector and slides of the proposal; or with PowerPoint and a laptop projector. The good part of the long form proposal, or even proposal sandwich, is that it allows one page to be reviewed at a time. It also allows for lots of questions and discussion before ever getting to the price. Again, when the price is shown, selling stops, or at least hesitates.

There is a little showmanship and "controllmanship" involved in this. There may even be a little drama. This is also called selling. Simply handing a proposal to a customer without control, or emailing or faxing, is called quoting or emailing or faxing or handing out. It's sure not selling!

Almost reluctantly, at the end of the proposal review meeting, the salesperson entrusts the proposal with a designated customer contact.

And, of course, asks for the order.

Handling situations where the customer will not schedule a proposal review meeting

In these situations, one alternative to emailing or faxing is to send the proposal via FedEx or other courier service. This has several advantages:

- It's more impressive when it shows up. This allows the salesperson control of the appearance of the document; it's still on his paper; it has tabs if needed; literature is intact; drawings can be included; signatures are originals, improving the personal nature of the communication.

Whether emailed, faxed, or sent by courier, the delivery of the proposal should almost immediately be followed by a phone call from the salesperson. The ostensible purpose of the call is to confirm the customer received it. The real purpose of the call, however, is for the salesman to ask the customer if he can "walk through the proposal with you over the phone." This is essentially proposal review by phone. While certainly not as effective as a personal proposal review meeting, it does allow the salesperson to ask questions and get direct feedback from the customer.

Simply calling to confirm the customer got the proposal and then stating, "If you have any questions, please feel free to call me" is a missed opportunity.

Another scenario sometimes encountered by the salesperson is the "lobby brush-off." This occurs when the salesperson shows up for a proposal review appointment and the customer comes to the lobby and says something like, "Something unexpected came up. Just give me the proposal and I'll review it and call you with questions." Before handing over the proposal, the salesperson should keep possession of the proposal (safely hidden in a briefcase or laptop is good) and ask for a rescheduled appointment. The face-to-face opportunity for proposal review is too important to give up easily. If forced to leave the proposal, the salesperson should try to schedule a phone call to "walk through the proposal" with the customer.

Summary

Again, the keys to effective proposal review with customers are:

■ Face-to-face is best

■ As many decision influencers present as possible

■ Control the document

■ Ask questions and listen to the answers

■ Ask more questions and listen

■ Use the **Objective Based Selling** proposal format as review/presentation outline

■ Work for trial close agreements on each page

■ It's not a sales pitch; it's a review of customer objectives, parameters, alternatives, recommendations, bill of material, investment, and proposal support

■ Try at least once to close the sale—even though in the significant project material handling sales environment, such closes are rare

Note: There is further information on both traditional and **Objective Based Selling** closing techniques in chapter twenty-two.

Okay, if the salesperson is unable to obtain the commitment (close the sale) in the proposal review meeting, he should prepare for further proposal support and look for the appropriate opportunity to modify the proposal and obtain the customer commitment (get the order).

Chapter Nineteen

Provide Proposal Support

Proposal support is reinforcement that the salesperson's proposal will help the customer meet their objectives, within the physical parameters of the situation. It's "proof" the salesperson's proposal and follow-through will live up to expectations. Its purpose is to build trust in the salesperson and his proposal, and to provide the customer the conviction to act on the proposal. In some situations, proposal support is also used to overcome specific, stated customer objections.

There are two basic time frames and types of proposal support:

■ Documentation or activities that accompany the sales proposal itself; in many cases becoming an integral part of the proposal

■ Documentation or activities that are provided or occur following the proposal presentation to the customer

In both cases—but more often in the second post-proposal situation—proposal support can be a catalyst for a trial close; in best-case scenarios, a sales "closer."

OBJECTIVE BASED SELLING

Proposal support accompanying proposal documents and reviews

Most often this is documentation. Examples include:

- Layout drawings of conveyor, racking, shelving, modular drawers, mezzanines, work stations, material flow, casters, etc.

- Insurance certificates (which are now required by most customers to work on their premises)

- Brochures of models proposed

- Performance charts of equipment proposed

- Warranty statements

- Organization charts of the salesperson's company (providing assurance and comfort that the organization is there to do what the proposal proposes)

- Capabilities statements of salesperson's company, suppliers, subcontractors (includes listing of specialty equipment, facilities, etc.)

- Listing of awards achieved by salesperson's company
- Time line for delivery and/or completion of purchase or project
- Professional résumés of key employees involved in implementation of project
- Pictures of facilities that will be involved in completion of project
- Copies of pertinent or specialized business licenses involved in the project: for example, contractor's licenses where applicable; stamps of professional engineers involved
- Where more than one supplier is involved in the implementation of a project, list all with their areas of responsibility and key contact people

There are many other possibilities. The idea is to choose proposal support that is appropriate to the specific customer and project or purchase under consideration—not an endless supply of meaningless documentation. The criteria for inclusion: Does this support indicate how this proposal will meet the customer's objectives in this situation? Or does the proposal support help overcome a customer's objection?

Two forms of proposal support should be included with every proposal of significance:

- List of companies for whom the salesperson's company has done similar projects successfully
- If available, letters of reference from companies for whom the salesperson has satisfactorily completed similar projects (the "Follow Through" chapter of this book will instruct when and how to obtain these letters)

The level of detail and format of the company or project lists can vary. These lists should be for projects completed within a reasonable past time frame, usually no more than three years. They can be a simple list, or contain some project detail. They should only contain individuals' names and contact information if this has been pre-approved recently with those persons, along with assurance they will be open to receiving reference checking calls or emails—and, of course, that the reference information will still be positive! This same reconfirmation and assurance is important also before reusing letters of reference.

Project lists and letters of reference can be the most powerful proposal support available. They indicate others have used your company in similar situations, and were satisfied.

In addition to documentation, the following are sometimes used at personal proposal presentations:

■ CD's, DVD's, or videos of equipment in action

Before using these materials, the salesperson must preview them from the perspective of their relevance and interest to this customer. Also, attention spans can be short for this type of material in personal presentations. Even three minutes can be too long. The salesperson should be prepared to show only relevant portions, not just "plug and play."

■ Actual equipment demonstrations

Demonstrations are among the most costly, poorly executed, and failed sales techniques used by material handling salespeople. It is not recommended these be used as proposal support at time of proposal review. This is because they are time-consuming, difficult to arrange, may distract from the proposal itself, and often involve unexpected experiences. Demonstrations are better performed in a more controlled environment, before final proposal presentation or as a trial close after proposal presentation. Guidelines for effective **Objective Based Selling** demonstrations are discussed in the next section of this chapter.

Proposal support following proposal presentation

As mentioned in an early chapter, a characteristic of the material handling selling environment is that decisions are not usually made at the time of the original proposal or even when the salesperson is in the room (nor when the salesperson is in telephone conversation with the customer).

So, if in spite of strong pre-proposal sales activities, an on-target **Objective Based Selling** proposal with clear alternatives at different price points, personal presentation in a scrum meeting, and good proposal support documentation accompanying the proposal—the sale is not closed at time of proposal presentation—then what?

It's time for post-proposal support.

The simplest version of this can be documentation not included with the proposal. In fact, an effective sales technique is to understand that the order most likely will *not* be closed at time of proposal presentation, and to intentionally leave some support out to use as a purpose for post-presentation contact.

This can be used as a reason to schedule a meeting with a key contact after the proposal presentation. At that time, some follow-up open-ended questions are appropriate:

> **"What was the group's reaction to our proposal? What was your reaction to our proposal?"**

> **"What should we be doing next to help with your decision-making process?"**

> **"What are your hesitations with our proposal?"**

And, the questions which should be asked throughout the **Objective Based Selling** process:

> **"What's changed since we last talked? Since our proposal presentation?"**

It's certainly appropriate in these post-proposal meetings to "ask for the order"!

If a personal meeting with key contacts is impractical or inappropriate following proposal presentation, additional proposal support can be mailed, emailed, sent FedEx, by courier... whatever it takes to get it there to have another customer contact.

Objective Based Selling demonstrations and site visits

In addition to documentation, two common post-proposal activities can be effective in convincing the customer that the salesperson's proposal should be acted on:

- Site visit to a facility where the salesperson has supplied equipment or completed a project similar to that being proposed

■ Demonstration of equipment the same as or similar to that being proposed

These are often expensive and time-consuming to make happen, so they should only be used where the size of the proposal or overall opportunity with the customer justifies it and where the site visit or demonstration will clearly show the customer how the proposal will meet their objectives or overcome an uncertainty or objection the customer has with the salesperson's proposal.

Overall guidelines for effective management of demonstrations and site visits include:

■ Remember, the salesperson's goal is not to put on a show, but to close the deal or move it closer to close by showing the customer how the salesperson's proposal will help the customer achieve their objectives by acting on the proposal. Secondary goals may include overcoming a specific, stated objection; acting as a catalyst to meet more decision influencers; determining the customer's true reaction to the proposal.

■ Establish specific customer agreed-upon objectives for the demonstration or site visit—objectives which the salesperson knows are attainable in the demonstration. For example, a valid goal of a demonstration might be to illustrate the forklift will operate in the customer's aisles, and handle their ramp. This can be shown fairly easily in a controlled demonstration. All will know at the end of the demonstration whether or not it was successful. A non-valid goal of a demonstration would be to show product reliability. That can only be demonstrated over years of use, not in a controlled, time-limited demonstration. Proposal support for that goal would be customer reference (one who has used the product over a period of time) or warranty assurance.

Establishing mutually agreed-upon (between salesperson and customer) objectives for the demonstration or site visit is the most critical guideline—and the most neglected. These objectives should be stated in writing prior to the demonstration or site visit. These are what make them **Objective Based Selling** demonstrations or site visits, instead of time-wasting, order-losing activities!

- ■ Approval of any site visit must be obtained from the facility and operational manger to be visited. The site should be pre-visited by the salesperson.

- ■ The salesperson should attempt to have as many of the decision influencers as possible available for the demonstration or site visit—at least for the beginning and end.

- ■ Equipment demonstrated should have the exact specifications of the equipment being proposed. Any variation is an invitation for a failed demonstration. "I know this doesn't work exactly right, but yours will be different and it will work." Right. So, why did we demonstrate in the first place?

- ■ The salesperson should always be present at the start and end of the demonstration or site visit, and preferably throughout. Following this guideline will, of itself, limit the length of demonstrations. The reason for this guideline is so the salesperson can "set the stage" for the demonstration (by repeating the goals and gaining reconfirmation); answer questions; deal with unexpected occurrences; hear stated objections and deal with them; "wrap up" the demonstration or site visit with a summary and agreement that the demonstration or site visit met stated goals. The salesperson will also be available to close the deal!

Specifically, at the end of the demonstration or site visit, the salesperson should restate the objectives and seek agreement that the objectives have been met (or objections satisfactorily dealt with). If not, revisit the issues or revise the recommendations proposed, based on new information gained. If there is agreement that the objectives have been met, ask for the order! If they are unprepared to give the order, ask these questions:

"What's your hesitation?"

"What other objectives or objections have we not dealt with?"

"What else do we need to do to earn your business?"

"Where does our proposal come up short?"

Proposal support should continue until the order is obtained, the order is definitely lost, or the customer indicates it's time to modify the proposal—which is the next phase of the **Objective Based Selling** process, and the next chapter.

Chapter Twenty

Modify Proposal

After receiving proposals and nearing the decision to purchase, customers often reexamine and double-check details. Things like objectives and their priorities; parameters; budgets; requirements; and differences between proposed alternatives. Considerable time may have elapsed between original conceptualization of the project and time for a decision; and as time passes, "things change."

New decision influencers enter the process (with new priorities, new information) and some original influencers may no longer be involved. The proposals themselves may trigger new ideas. Alternatives and information in the salesperson's proposal may suggest new or "fine tuned" requirements.

In the author's experience, over 80 percent of initial proposals for significant projects are subjected to customer requests for "modified proposals." In many cases, there will be requests for more than one modification.

The salesperson should understand the request is not uncommon, and as indicated on the **Objective Based Selling** diagram, is an expected part of the sales process. The job of the salesperson is to figure out what is really causing the request for proposal modification, so he can determine the most effective response and the appropriate effort to put into this modification.

When customers request modified or revised proposals, there are

OBJECTIVE BASED SELLING

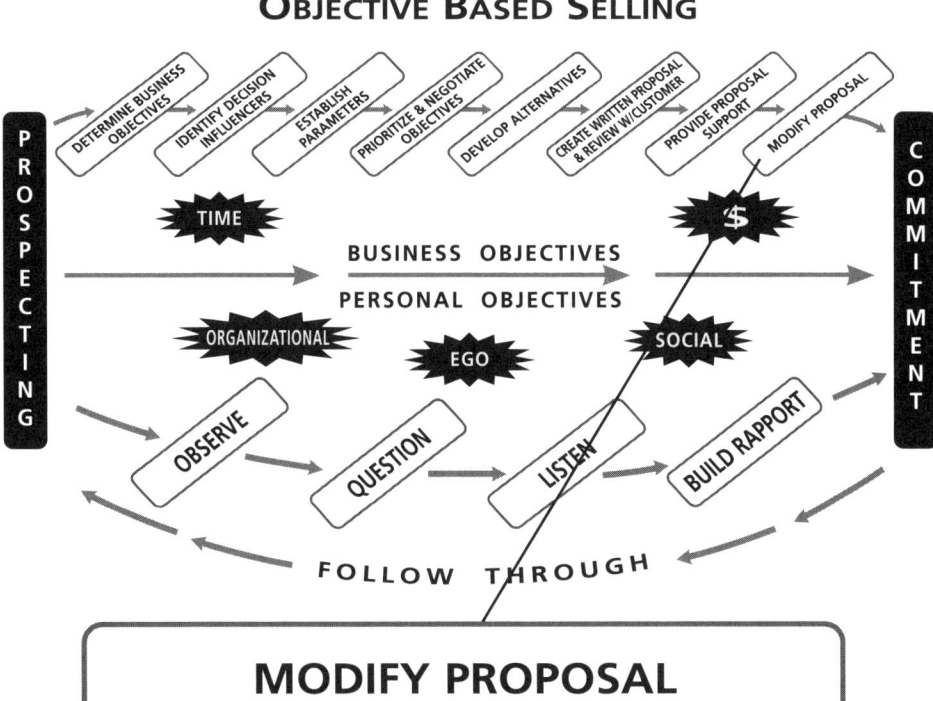

MODIFY PROPOSAL

four possibilities:

■ The salesperson is being subjected to busy work requests while the customer tries to figure out what is really required or frantically tries to meet the requests of a "higher up" decision influencer.

■ The customer has decided to buy from someone else, and is simply using the salesperson to keep the other preferred supplier honest.

■ The customer is really zeroing in on their requirements, and the sales process must begin again.

■ The customer is giving the salesperson from whom he *really* wants to buy the information to finalize a proposal that will be accepted.

The need to modify the proposal often occurs near the end of the decision-making process "at the last minute" as the customer tweaks the details, quantities, specifications, etc. While the decision-making process may have seemed leisurely up to this point, the project champion now may seem in a hurry to get the revised proposal—often while he still has the approval person's attention. Unless a public bid situation, the customer is under no legal or ethical obligation to call all potential suppliers to ask for modified proposals. Often they will call just the one they trust the most—the supplier they have already decided to do business with if things go well from here. Getting this second chance or call to modify the proposal can be a true buying signal, but in spite of the urgency, the salesperson should move carefully to be sure of the situation and to avoid wasted effort or misstep at this point.

When given a request or an opportunity to submit a modified proposal, the salesperson should thank the customer, then ask open-ended questions to determine the current situation. These questions include:

"What triggers this request?"

"What has changed?"

"Other than these changes, what was the reaction to our original proposal?"

There should also be questions about the process going forward:

"Who will review the revised proposal?"

"What will be your process for reaching a decision following this revision?"

"What is your time frame for a decision following proposal re-submittal?"

"What is the urgency for the revised proposal?"

The salesperson must be listening carefully to the customer's responses—and asking himself the implications.

The most desirable possibility for a salesperson is that the customer has decided he wants to do business with the salesperson, but knows that the original proposal—for whatever reason—won't be acceptable. Perhaps a competitor has proposed a different model which will work; perhaps pricing is out of line; perhaps it's as simple as a change in quantity or a revised layout based on new information. This type of modified proposal is also called "the second chance." The salesperson has earned the right to do business, and now must get the details right!!! Often in this situation the customer will coach the salesperson he wants to do business with as to what is required to achieve an actionable proposal.

If the salesperson feels this is not the case, he now must decide if he still has a fighting chance for the order, and if so, what portion of the **Objective Based Selling** process must be repeated in order to be successful:

Are there new decision influencers to know?

New objectives or priorities to uncover?

Objections that are surfacing that require more proposal support?

Another whole proposal presentation to try for?

New parameters?

Another alternative to suggest?

The salesperson has the **Objective Based Selling** model as a guide to reestablish control of the sales process—or to decide it can't be done, and to move on to other projects.

Closing the sale

While decisions on significant projects are most often made when the salesperson is not in the room, the Objective Based Selling model provides many opportunities for trial closes. Certainly, providing the customer with a revised or modified proposal is one of those opportunities:

"May we process your order based on this revised proposal?"

Additional traditional and **Objective Based Selling** closing techniques and questions are discussed in chapter twenty-two.

Chapter Twenty-one

Build Personal, Professional Relationships

Many of the earlier chapters of this book have dealt with the business side of the **Objective Based Selling** loop. This is the structured, organizational, formal process of determining business objectives and operational parameters: meeting decision influencers; offering alternatives; preparing and reviewing proposals; and providing proposal support and modifying proposals. It involves the customer's formal decision-making and purchasing processes.

However, despite much of the depersonalizing processes and forces of the business world—voice mail, email, restricted access, websites, formal criteria, committees, budgeting processes, national agreements—in significant material handling purchases, people still buy from people.

Where national contracts are set up, personal relationships help determine who they are set up with; how well they are enforced; and how they can be circumvented.

Where access is restricted by phone systems, electronic supplier registrations and email, customers give access to whoever they want

OBJECTIVE BASED SELLING

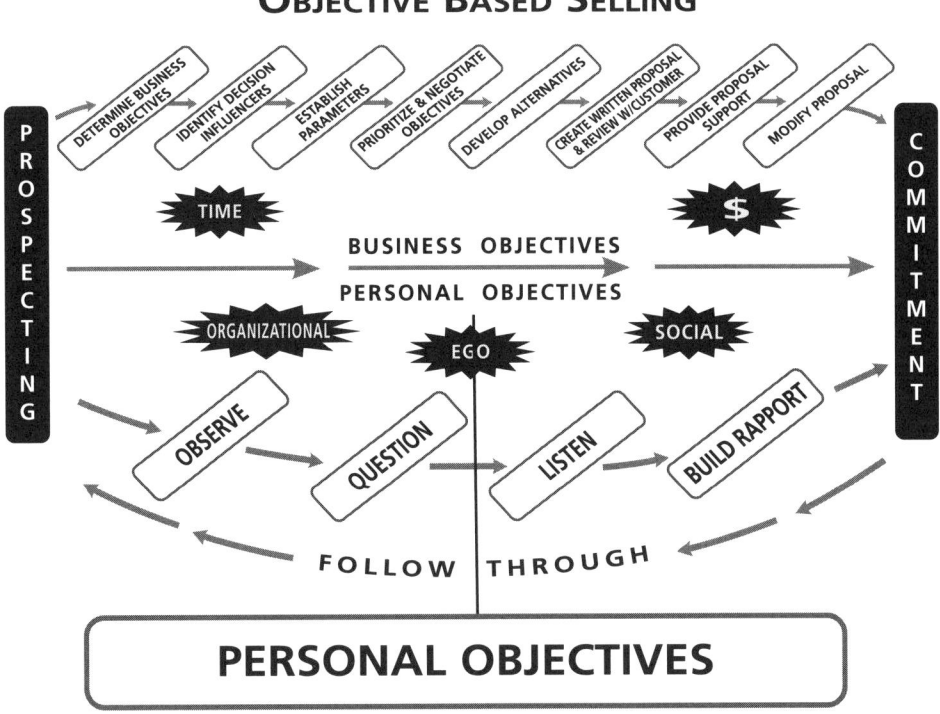

to do business with.

In situations necessitating urgent or last-minute requirements, customers contact salespeople and companies they like and trust.

Where tight purchasing systems have been created, company and even individual credit cards are used to bypass them. Even federal government—including Department of Defense and GSA purchasing systems—have loopholes through which individuals make significant purchases daily.

Contract bidding systems are routinely "worked" to award bids to contractors and subcontractors individually selected, often for personal reasons.

Attempts have been made by many customers to treat some material handling equipment like commodities, including the use of reverse auctions. In most cases involving significant purchases of material handling equipment, where reverse auctions have been tried, they have been abandoned due to the customized application nature of material handling purchases and the importance of this equipment and services to customer operations. Some parts and minor, standardized items are purchased online without salesperson involvement. However, personal sales relationships are often involved with these systems, in directing customers there and resolving issues which may arise.

As articulated by Ron Willingham, consultant, people buy from people they:

■ Like

■ Believe

■ Understand

■ Trust

Concurrent with the business side of the Objective Based Selling loop

Simultaneously with working the business side of the **Objective Based Selling** loop, material handling salespeople must be working to understand the personal objectives of the key customer contact and decision influencers. This is accomplished using interpersonal skills along with business skills and systems. It is done by observing, asking questions, listening, and building rapport with these key individuals. While a less formalized process than the business side of the loop, it is no less important. It is, in fact, more important.

If customers don't like, believe, understand, or trust a salesperson's proposing significant material handling products, equipment, services, or changes, they will find a reason and a way to buy from someone they do.

They will seldom tell the material handling salesperson who is losing the order that that is the reason. They may not even admit it to themselves.

This chapter will explore this subject and offer effective methods for salespeople to build personal, professional relationships, in four ways:

1. Discussion of what makes customers like, believe,

understand, and trust salespeople in the material handling sales environment

2. A look at appropriate observations and questions for the personal side; what to listen for and how to build rapport

3. Identifying two basic customer types at opposite ends of a relationship continuum—and how to deal with them

4. Dealing with the ultimate customer objective: their ego!

1. Customers must like, believe, understand, and trust you

Following is an examination of the kinds of activities and behaviors most likely to help salespeople achieve these feelings in the business-to-business sales environment. These are, of course, generalizations. There are always exceptions.

Customers will **like** salespeople when salespeople:

- Show genuine interest in them (asking open-ended questions and listening, instead of just talking)
- Have an appearance that seems appropriate—to the customer
- Are interested in some of the same things
- Help make their job easier
- Have speech patterns and activity paces similar to theirs
- Spend non-pressure time together
- Don't talk down to them
- Share something with them: an activity, lunch etc.
- Make customers look good for their organization and boss
- Conduct themselves with appropriate social conventions
- Have an appropriate sense of humor

And, well, sometimes it just clicks!

Customers will **believe** salespeople when salespeople:

- Research before making claims or statements or recommendations
- Make claims that are confirmed by their own experience or by others in their organization
- Don't bad-mouth competition
- Show proof statements or evidence
- Are detail oriented

Customers **understand** salespeople when salespeople:

■ Use words customers understand (not jargon they don't
   understand)
■ Walk customers through a discussion step-by-step without
   talking down to them
■ Create opportunities for customers to see it with their own
   eyes
■ Support statements with appropriate visual aids

Customers **trust** salespeople when salespeople:

■ Show up on time
■ Give a strong hand shake, and look customers in the eye
■ Meet commitments—even small ones, and especially
   important ones
■ Don't exaggerate
■ Make the customer look good for their organization and boss
■ Resolve problems quickly
■ Always follow through

These are not meant to be exhaustive lists, just excellent guidelines for behavior to build personal, professional relationships.

2. Observe, question, listen, build rapport

Salespeople should be observing many things with customers,
beginning even before the initial contact:

■ What can I learn from their website? Facility appearance?
■ What's the tone of the lobby of the customer: Friendly?
   Professional? Impersonal?
■ How protected or isolated are primary contacts and decision
   influencers?

During initial interviews, salespeople should observe:

■ Is the customer on time, and respectful of the salesperson's
   time?
■ Does customer give salesperson undivided attention?
■ What is in the contact's office? What does he seem interested
   in? What seems to be the appropriate attire? How organized
   does he appear?

- How rushed is he? Where is his office in the facility? In the apparent pecking order?
- Is he detail-oriented? Interested in the bigger picture? Both?
- What seems to be his primary method of communication (phone, Blackberry, email)?
- What's his sense of humor?

During facility tours, the salesperson should observe:

- How does he appear to treat others in the organization? How does he appear to be treated by others?
- What in the facility interests him?
- What areas does he appear to have responsibility for?
- What's the condition of the facility? Demeanor of employees?

In general, the salesperson should be observing the best way to relate to the individual, and how he seems to interface with his own organization. The salesperson should also be observing if it appears the customer's way of doing business will, in the long term, be compatible with the salesperson and his company's way of doing business.

Regarding questions, other than the business objective questions discussed in other parts of this book, salespeople should consider relatively personal open-ended questions such as:

"How long have you been with this company?"
"What's it like to work here?"
"What do you like about working here? About your job?"
"What's the most difficult part of your job? Of working here?"
"What's been your favorite job? Why?"
"What kinds of things do you do for fun?"
"What do you do when you're not working?"
"Where are you from?"
"Where'd you go to school?"

If the salesperson sees a personal item, or item indicating outside interests, or awards, these can be appropriate questions:

"Tell me about that (picture) (award) (plaque) (item)."

The salesperson must constantly be sensitive to the customer's desire or willingness to discuss these things.

On more professional topics, questions might include:

"What is your area of responsibility?"
"What other material handling projects have you been involved with?"
"What has your experience been with our company?"

The general idea of personal, professional questions is to find common ground; mutual interests; and most importantly, find out how the customer wants to do business. The salespeople who are most successful are those who do business the way the customer wants to do business. Personally and professionally. If there is a basic incompatibility in the way the customer does business and the way the salesperson or his company does business, the salesperson must evaluate whether a successful adaptation can be made or whether he should move on to another customer.

As to listening on the personal side of the **Objective Based Selling** loop, the salesperson is listening for clues as to how the customer wants to do business. Not only are the words important, but the customer's tone, style, implications, and interest. The salesperson is listening for clues as to the best way to relate to the customer and build personal, professional relationships.

Build rapport

Most successful salespeople will develop a true friendship with one or a small number of customers, sometime during their career. These close relationships will often include spending significant time outside a strict business environment. It may involve a common sporting interest like golf or fishing; it may involve an outside organization both are passionate about.

However, this is not the general goal for salespeople when developing personal, professional relationships. Friendship is too rare and cannot be planned. However, salespeople can have a goal of developing rapport with key decision influencers at customers they target for business. The dictionary speaks of rapport as a relationship marked by harmony, conformity, accord, or affinity. That's the goal of observing, questioning, listening: determining how best to achieve this level of relationship. If it can't be done, significant business will most likely not be achieved with this customer.

3. Personal or professional first? Two customer types

While generalizations about people are always suspect, there are two basic extremes of customer individual types which are helpful to examine.

At one extreme are customers who are not interested in making a personal connection of any significance until the salesperson has proven himself professionally. In other words, the salesperson must earn the right to have a personal conversation or relationship. A salesperson trying for a personal conversation in an initial interview with this type will be rebuffed. If the attempt is continued, the salesperson will be seen as insincere and further access will be difficult. Signs of this type for the salesperson to observe can include an office without many personal items; a somewhat brusque, get-to-the-point manner. Short or non-answers to personal questions is another tip-off, as well as short allowances for appointment time.

At the other end of the personal–professional style continuum are individuals who want to make a personal connection first, before doing significant business. They want to get to know the salesperson as a person, before deciding if they trust them enough to do business. Indications of this style may be an office with lots of personal memorabilia; expanding on answers to questions of a personal nature; asking some personal questions of the salesperson, and looking for connections. Salespeople dealing with this type must slow down and look for a personal connection. It may take time, and the salesperson will have to determine if the eventual payoff is worth the effort. A particular challenge with this type customer is they may already have personal, professional relationships in the area of the salesperson's business responsibility, and competing with that is difficult. One strategy may be to find a different decision influencer in the customer's organization to connect with first. Another strategy is to find someone in the salesperson's organization who may be able to make a better personal connection. A third strategy is... patience.

Most customer contacts are not at the extremes of these types, but somewhere on the continuum between professional first, then personal; and personal first, then professional. Assessing this and acting accordingly will help the salesperson build rapport.

4. Ego

Just as every viable material handling project has time and money objectives in addition to others, every individual has ego. Their self-concept. How individuals view themselves impacts their self-confidence, their style in dealing with others. Like salespeople.

Again, two extremes might be individuals who have strong self-concept, sometimes to the point of arrogance, and others who have low self-concept, to the point of low self-esteem.

There are no easy open-ended questions to ask to determine the self-concept of an individual. You can't ask, "How's your ego: high or low?"

Salespeople will, however, always be aware that they will be most successful in those situations where they make an accurate assessment of this and treat customers accordingly. The complexities of this are beyond the scope of this book. However, there is one book which has taught effective ego-handling techniques for over seventy years.

All salespeople are encouraged to read the book *How to Win Friends and Influence People* by Dale Carnegie. It is available in updated form in most bookstores, usually in paperback. Buy it. Read it. Practice it.

Handwritten notes and thank you cards

In a world of email, personal digital assistants, instant messaging, and texting, a somewhat forgotten, neglected, abandoned way of connecting personally with another person is the handwritten letter, note, or personally signed card.

Because this method of communication has largely fallen into disuse, it can be an effective way for a salesperson to connect personally with some decision influencers or gatekeepers—a way to stand out from the crowd.

There is something satisfying about seeing your name handwritten by someone and there is a unique connection to a personal signature.

It can almost be guaranteed that if a person is sorting a pile of mail, among the first things to be opened will be an envelope hand-addressed personally to them, with a personal return address written on the envelope.

Some sales companies supply "thank you" or other cards for their salespeople to use, usually with the company logo preprinted on the cards and envelopes. These are good to use, but only if the salesperson hand addresses the envelopes and hand writes the note inside the card.

If the salesperson buys his own note cards (visit a good stationery store; packs of cards are inexpensive and worth the investment) and writes the message, along with hand addressing the envelope, it can have an even more "stand out" impact on the customer or prospect's desk. Use of a real postage stamp instead of metering almost guarantees the envelope will be opened. This says, "You're important enough for me to hand write this and buy my own stamp."

It certainly is a positive thing for most people's ego!

What to say on the card or what to thank the customer or prospect for? It can be almost anything. Thank them for taking your call; for giving you an appointment; for calling you back; for giving you an opportunity to submit a proposal; for introducing you to someone else; for a referral; for attending an event. Of course, write them a personal note thanking them for orders. And, if the customer gives the order to someone else, imagine their surprise if they get a thank you card for providing you the opportunity and for giving feedback on your proposal.

Thank you notes to receptionists and gatekeepers go a long way to getting you remembered and for receiving special treatment on the next visit or contact attempt.

In addition to "Thank you," "Congratulations" is another good reason to write a personal note or card. Tear out positive articles about the contact or their company and send with a personal note congratulating them. "Congratulations on the recent achievement; new plant; award; newspaper article, (_____)."

Sending computer-labeled holiday cards imprinted with the company name inside and run through the company postage meter doesn't really say "You're important—I'm wishing you the best at this holiday season". Personally addressed and signed cards, with real postage stamps, do.

One goal for every material handling salesperson: Send at least one personal, handwritten note or card every day—to a prospect, customer, or gatekeeper. Every day.

Business and personal coming together

As the **Objective Based Selling** diagram indicates, to achieve significant material handling sales, salespeople must work on the business and personal sides of the **Objective Based Selling** loop simultaneously. Specifically, salespeople must work to uncover business and personal objectives and show the customer how they can achieve these objectives by dealing with and acting on the salesperson's proposal.

These seldom happen at the same pace. In some situations, salespeople will make an immediate personal connection with key decision influencers, and this will facilitate their work with the customer business processes. In others, by working the business side diligently, a personal relationship will form.

Whatever the pace, the salesperson will consistently achieve significant business only when these come together at the time of the customer's commitment.

Chapter Twenty-two

Obtain Commitment

"Closing the sale" techniques are taught in many sales seminars, tapes, CD's, books, and coaching sessions. Traditional techniques commonly taught include:

- Trial closes. This is often taught as getting the customer to agree to do less significant, non-buying tasks; or to agree to simple statements escalating in importance until the decision to "go ahead," or "go with you," or to purchase is simply another seemingly inevitable "yes" in a series of decisions. Trial closes in **Objective Based Selling** include confirming key objectives and parameters; agreeing to introduce the salesperson to other decision influencers; providing credit information or filling out a credit application; committing time to give the salesperson a facility tour; helping put together a scrum meeting; considering alternatives; attending a proposal review meeting; demonstrations or site visits.

- The "old standard" choice close: "Would you prefer the new truck, or the used one?" "Would you rather pay cash or finance?" and so on.

OBJECTIVE BASED SELLING

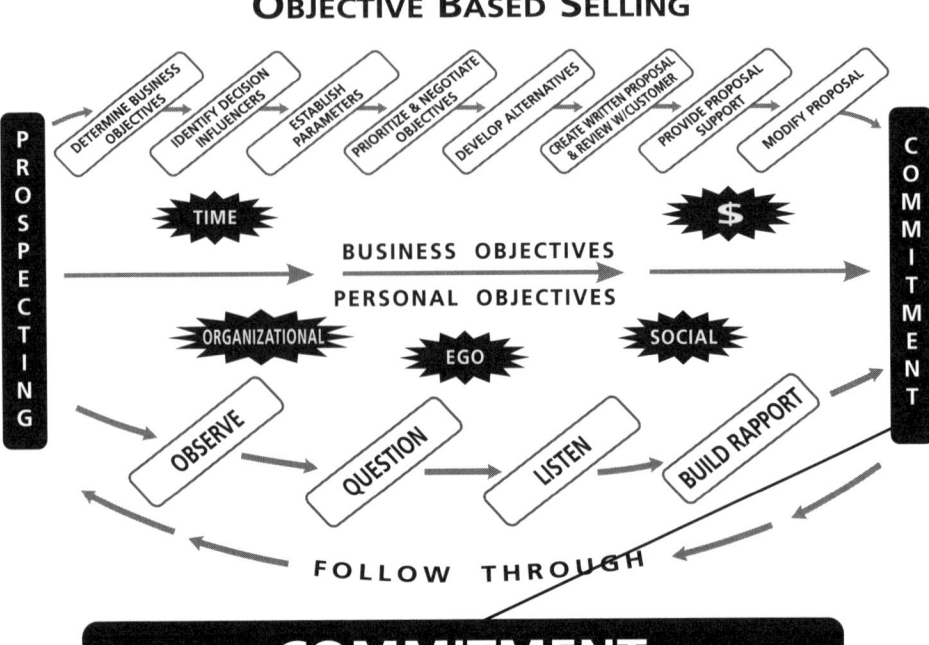

- "Processing the paperwork" close: "Can we get the paperwork started?" The implication is you're not really buying anything, we're just processing necessary paperwork.

- The assumptive close: just moving forward with project details, delivery, paperwork, on the assumption the customer is buying from you—unless the customer steps in and says, "Stop."

- "Ben Franklin" close: dividing a sheet of paper in half and listing the features and benefits of the salesperson's proposal or equipment on one side, and the competitor's on the other (guess whose list is longer and more convincing?)

- Equipment salespeople love to use the "The price is going up, better buy now" close, or the "To meet your delivery, we better order now" close. (Caution: Customers often don't believe this!)

- The "puppy dog" close is sometimes used with equipment that can easily be delivered: "Let's just bring one out and if you like it, we'll leave it and send you the invoice."

■ Simply asking for the order: "May I put this on order for you?"

There are seemingly endless variations on these and other closing techniques.

How should these techniques be used by material handling salespeople in the Objective Based Selling process?

Whenever a material handling salesperson finds himself in a "classic closing situation," where...

■ the primary decision influencer or maker is present in person or over the phone;

■ objectives and parameters have been established, prioritized, are fairly straightforward, and are met by the product or service offered by the salespersons;

■ dollar or method changes by the customer are not significant;

■ and the customer needs to make a decision soon due to his time frame objectives;

...the material handling salesperson should draw on one of the classic closing techniques, and try to get the order. If it can be done, great. Move to order processing and follow through.

In other words, any time it appears it might work, try for the traditional close—particularly if you've earned the order by working with the customer through their objectives and parameters.

However, two distinguishing characteristics of the material handling sales environment limit the effectiveness of these traditional closing techniques. One is that due to the significance of many material handling sales and the number of decision influencers involved, purchasing decisions are most often made when salespersons are not in the room. Yet, many traditional closing techniques are based on the salesperson being in direct, personal contact with the customer—either physically with him or talking on the phone. If this is not the situation—and it is not the situation in significant material handling sales—these closing techniques don't work effectively, if at all.

With multiple decision influencers involved with purchasing in larger organizations, more than one "sign-off" or approval is often required: sometimes in a specific sequence. Getting the agreement of one decision influencer in a classic closing situation is only the first step of many; and again, the salesperson most likely will not be present for later steps in the approval process.

Another challenge for traditional closing techniques in material handling is the issue and process of modifying proposals. The salesperson may be in a proposal review meeting with all the right people and they may like the proposal and respond to a closing technique. Even then, when proposals must be modified as a result of new facts or input from that meeting, the tentative agreement to purchase may change with changes in the proposal.

So what's a salesperson to do? What is the closing process in this selling environment where "traditional closing techniques" may not work?

It's a commitment

To be effective in closing sales in the material handling sales environment, the salesperson must first understand that in any significant purchase or project, the decision to go with the salesperson's proposal is more than a simple purchase: It's a commitment. Because the dollars are often significant, and the purchase may involve a whole new way of dealing with material handling situations, the purchase also involves committing the buyer's organization to the seller's organization, to their way of doing business, follow-through, and capabilities.

Understanding that commitments are a deeper level of decision than a simple "purchase" is a first step to avoiding the frustration of looking for chances to try traditional closing techniques—and not finding them.

With the selling environment for significant material handling projects, the act of earning the order is often not a simple sales trick at the moment of decision. Rather, the customer commitment is the logical extension of a process, the process outlined in the **Objective Based Selling** model.

During this process, key customer decision influencers have

come to trust the salesperson's understanding of their situation, and thus his recommendations. The personal relationships developed provide the assurance for the customer that the decision to go with the salesperson's proposal (not just his product or service but with his proposal as a way of doing business) will be the right one for the customer. The customer trusts the follow-through to make the decision work to their expectations.

The salesperson does not have to be in the room when the decision is made to purchase from him, because:

- The salesperson has taken the time to understand the customer's business and personal objectives and prepare a proposal that clearly outlines them
- The salesperson provides responsive recommendations with alternatives
- The salesperson has provided appropriate support for the proposal
- The salesperson has personal professional relationships which promote trust, and provide internal advocacy for the salesperson's proposal inside the customer organization.

Trial closes and buying signals in Objective Based Selling

As mentioned above, the **Objective Based Selling** process offers many opportunities for two types of trial closes:

- customer agreement on something other than the actual decision to buy
- customer does something proactive at the request of the salesperson, indicating growing confidence in the salesperson and his proposal

Agreement trial closes can occur during identification of objectives and parameters; prioritizing objectives; considering alternatives. Even calling a salesperson back and agreeing to meet for an initial or follow-up interview with the salesperson is a trial close.

Even more positive trial closes occur when the customer takes proactive steps to work with a salesperson. Examples include supplying extensive technical information (usually involving parameters) to a salesperson; introducing the salesperson to other decision influencers; supplying a reference or referral; helping create or attend a scrum meeting; allowing a salesperson to personally present his proposal in a review or scrum meeting; allowing an **Objective Based Selling** demonstration; going on a site visit; visiting a supplier's facility; and asking the salesperson to provide a modified proposal under the right circumstances. All of these activities are decisions by the customer to spend more time and in some cases internal political capital with the salesperson—and all are buying signals. The salesperson's request for these activities is the trial close.

Alternatively, reluctance or refusal to do these things, or suddenly not returning calls, are definitely bad buying signals!

Open-ended questions to be used when the buying signals stop in response to trial closes include:

"What's your hesitation?"

"What's changed?"

"Where am I coming up short of your expectations?"

"What should I be doing to work more effectively with you?"

"What should I be doing to help you implement this project?"

Objective Based Selling proposals help gain commitment when the salesperson can't be there

When traditional closing situations don't exist—drastically reducing the opportunity and effectiveness of traditional closing techniques—the most effective single closing tool to gain customer commitment is the **Objective Based Selling** proposal.

These speak to the decision influencers the salesperson can't meet; these speak in the meetings the salesperson can't attend.

The proposal becomes a primary differentiator from competitors; it is the catalyst for customer decisions and action. The **Objective Based Selling** proposal starts with customer objectives and parameters and builds a bridge to the salesperson's recommendations, or alternative recommendations. It provides proposal support, including references.

The proposal, along with proposal review scrum meetings where attainable, gains commitment and closes sales even when the salesperson isn't there.

And, again using the **Objective Based Selling** model and process, if there is a delay in ordering, proposal support extends contact with the customer, possibly creating a classic closing opportunity or a proposal modification opportunity which leads to the close.

The reverse time line: a special closing technique

In many material handling projects, use of a reverse time line can be an effective closing technique. It can work for the salesperson even when he is not present when key decision influencers meet, or when they make decisions.

In most material handling projects, particularly those of significance, at some point in the project a time frame (perhaps an operational deadline) becomes critical and becomes a driving force for making project and purchasing decisions.

This is particularly true in projects involving construction deadlines; lease expiration dates, either for facilities or for equipment being leased and replaced; and budget spending deadlines. (Some budgeting processes require the money actually be spent before a certain date, or it's lost. And some procedures will not allow it to be spent until the equipment is delivered, installed, or otherwise operational.)

Customers routinely attempt to pressure salespeople with these time frames: "I'll give you the order if you can meet this deadline" (often, an unrealistic deadline, again often created by customer procrastination in making a decision).

Salespeople can anticipate this, and if they act early enough to put a convincing, honest reverse time line in writing, actually use it as a catalyst for the customer to act. The beauty of this technique is it starts with the customer's own deadline—their time objective. It then builds in the steps for effective order implementation, to reach a time at which the customer must place an order to meet their own deadline. (Review example on next page.)

Example of Reverse Time Line

Customer has a June 30 deadline to have a conveyor system operational.

A reverse time line might look like this:

June 30: conveyor system operational at specification
Allow one week for test:

June 23: conveyor must be installed and running
Allow three weeks for mechanical installation, and two weeks for electrical installation:

May 23: conveyor must be unloaded at job site
Allow one week and two days for freight to job site and unloading, check against bill of material:

May 14: conveyor must ship from factory
Allow five weeks for production:

April 7: order must be processed at factory
Allow three days for order entry and processing to factory:

April 4: drawings must be approved by customer
Allow one week for customer review of drawings and approval:

March 28: drawings must be ready for customer review
Allow one week for final drawings after order is given to salesperson:

March 21: firm order must be given to salesperson in order to meet customer operational date of June 30.

The level of detail here gives credibility to the information. It also reminds the customer of all the steps in meeting his objectives—it's a sequence chart. It gives the customer confidence that the salesperson has done this before and knows what it takes. And, if it is presented with the proposal on March 14, it immediately gives the customer pressure to make decisions. Why not go with the salesperson who knows what he's doing?

Of course, the customer may question certain detailed time frames: "We can approve those drawings in three days; we can help expedite freight. How can we help you do the installation more quickly?" Of course, these are all buying signals in response to the trial close of the reverse time line. And, the customer is now working with the salesperson on delivery issues, instead of challenging him.

Compare this approach to the normal material handling salesperson's approach to giving delivery information:

Delivery: 90 days to 120 days ARO (assumes the customer
    knows ARO means *after receipt of order*, and which is it—
    90 days or 120 days? Delivered or operational?).

Reverse time lines give the customer information; build confidence in the salesperson and his organization; work for the salesperson to close the order when he can't be in the room when the decision is made.

Summary

Due to the significance of the sale, ordering in material handling is really the act of making a commitment to the salesperson and his company.

Traditional closing techniques often don't work effectively because the salesperson is not in direct contact with the customer when the decision is made.

Traditional closing techniques should be tried where the right conditions exist: relatively simple decisions, decision influencer and/or decision maker present or in direct contact with salesperson at decision moment; time becoming urgent.

Trial closes should be used throughout the **Objective Based Selling** process, getting customer agreements and their proactive actions.

Committing to the salesperson (ordering) becomes the logical culmination of the confidence built in the **Objective Based Selling** process. The proposal is a major key to obtaining this commitment. It speaks to the decision influencers the salesperson can't meet; it sells in the meeting the salesperson can't attend.

Reverse time lines can be effective closing tools in many material handling sales situations where customer time objectives are critical or urgent.

Chapter Twenty-three

Follow Through

The material handling salesperson has two primary follow-through responsibilities at the completion of a purchase or project:

- Assure the customer is satisfied. This means obtaining the customer's agreement that the purchase or project meets mutually agreed-upon objectives.

- Facilitate payment in full.

This follow-through, while a responsibility, is also a major opportunity. As shown on the **Objective Based Selling** diagram, successful completion of follow-through is the first step in prospecting for new projects or purchases.

Assuring customer satisfaction

Assuring customer satisfaction is not a single event. The salesperson's responsibility is to monitor the progress of a purchase or project throughout its implementation. As mentioned early on, a function of the salesperson is to help deal with unexpected situations or problems that develop in a project. These may include early or late deliveries; changing objectives or parameters; unanticipated and unexpected facility situations; delivery problems; product changes not communicated clearly by the supplier; or simple mistakes by customer, salesperson, supplier, or salesperson's company or subcontractors.

OBJECTIVE BASED SELLING

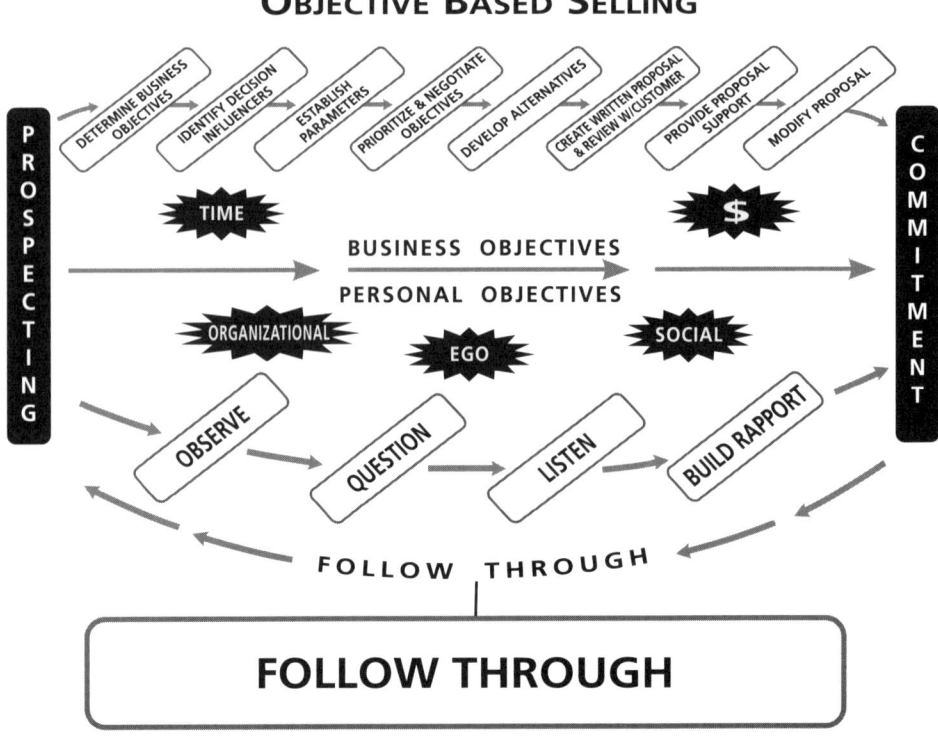

As these issues are identified, the salesperson should be the cat-alyst to organize resources to handle them. These steps are cus-tomer-, company-, and situation-specific, and details must be han-dled by the salesperson within their company's guidelines.

There are two general categories, however, to be dealt with. Some situations may call for a formal change order, including more money from the customer (particularly where objectives or parame-ters have changed—a major benefit of putting these in the proposal and indicating the product or service recommendation is based on these). Other triggers for requesting additional money from a cus-tomer can be changes in circumstances such as facility issues; reductions in quantity; last-minute options or alternative changes. The material handling salesperson's responsibility here is to pre-pare the proposal for the additional work or product change and investment; explain this to the customer and obtain approval; and make it happen.

If the change is not an issue requiring an additional customer charge, then the salesperson's responsibility is to explain this to their own company or supplier in a timely manner, and expedite corrective action.

When it appears a project is complete, the salesperson should ask the customer questions like:

"How have we met your expectations on this project?"

"How have we helped you meet your objectives on this project?"

"If you were doing the project again, what would you do differently?"

"What did you like best about the way we handled this project?"

"What should we have done differently?"

With customer agreement that the project has met expectations and objectives, the timing is right for the opportunity: the prospecting. Prospecting at this juncture is of two types:

- prospecting for another project with this customer—perhaps an "add on" triggered by the completion of this project (or a whole new purchase or project being planned)

- a referral to another contact in this company, or a referral outside the customer's organization

This prospecting, not surprisingly, takes the form of open-ended questions:

"What changes does this project implementation bring to your area which might trigger additional projects or purchases?"

"Who else in your organization is considering capital projects? Similar projects?"

"If you were me, who would you want to show this project to?"

"What other organizations do you know that might benefit from this type project?"

This is also the best time to ask the customer contact for a letter of recommendation, referral, or satisfaction. This is one time a closed-ended question might work best:

> **"Based on your satisfaction with this project or**
> **purchase, would you write a brief letter of**
> **recommendation or satisfaction with our work?"**

The salesperson can indicate it will be used appropriately to support other proposals for similar work. Since not everyone is comfortable writing letters, it can be appropriate to offer to write a simple letter, and ask them to put it on their stationery over their signature. It is important to get it on the customer organization's letterhead, if possible.

Some organizations have policies against this. As an alternative, the salesperson can ask if the customer's name can simply be listed as a satisfied user of his company's products and services. Another question here is to ask if the customer contact would accept a call from another prospective customer of the salesperson, to simply answer questions about their experience with the project.

A high percentage of really satisfied customers will agree to one of these satisfaction confirmation mechanisms.

Payment follow-through

A somewhat less pleasant part of the material handling salesperson's job is to help their company, as needed, obtain complete and final payment. In most situations, this is handled at the beginning of the project, by the salesperson performing their commercial transaction functions of getting the paperwork right; understanding customer payment practices; explaining their company terms to the customer; and listening to what the customer says is required for final payment.

In a small percentage of situations—often the larger, more complex projects—the salesperson must be involved in handling paperwork or other issues at final payment time. Delays in final payment may be caused by:

- Poor communication about completion status within the customer's organization

- Incorrect or unclear invoicing documents

- Changes in project scope or investment not clear between customer and salesperson's company

- Invoicing by method unacceptable to customer; for example, invoicing by paper when customer requires electronic invoicing

- Contract paperwork (e.g. lien waivers, customer forms) not complete

- Poor understanding by customer or salesperson's company of the other's terms or payment practices

- Customer not really satisfied; open issues to be dealt with

- Customer cash flow problems

While it is not the salesperson's role in most material handling companies to be a collection agent, it is in the salesperson's best interests to follow through as needed. If the customer is perceived as a poor payer, the salesperson's company may be reluctant to do another project with them.

On the other hand, if the salesperson's company is seen by the customer to be poorly administered in this area, or too difficult in collection, the customer may not want to do another project with the salesperson's company. And, in many material handling companies, the salesperson is not paid commission in full until the customer has paid in full.

Additionally, many of the issues listed above are most effectively handled by the person who knows the most about the customer and project and who has the personal relationships which can be most effective in solving payment issues: the salesperson!

Follow-through

Of course, the ultimate follow-through question is:

"What is our next project?"

This is the beginning of the **Objective Based Selling** for the next project!!! It closes the personal and business loops.

Chapter Twenty-four

Objective Based Selling: The Summary

Selling is an exciting, challenging, and rewarding profession—critical to our market-based economy. The material handling industry is essential to the smooth functioning of society.

Consistent use of a sales model provides salespeople the structure which allows effective use of individual sales skills.

Objective Based Selling is a business-to-business sales model to help material handling salespeople sell more at higher gross margins.

The **Objective Based Selling** model deals with the specific sales environment encountered in selling material handling equipment product and services.

Objective Based Selling focuses the salesperson and sales process on the customer instead of on the seller's equipment, services, company, and self. This model is designed for the salesperson to help the customer with project definition, information gathering, and decision-making functions—rather than to present a quote and try to close the deal.

A basic tenet of **Objective Based Selling** is that customers make buying decisions for business and personal reasons: to accomplish business and personal objectives.

The job of the salesperson is to determine the customer's business and personal objectives and show how they can best achieve those objectives by acting on the salesperson's customer-focused proposal.

Four keys to **Objective Based Selling** are:

- **Open-ended questions**—focusing on the customer, encouraging the customer to talk

- **Personal, professional relationships**—people buy from people they like, believe, understand, and trust

- **Customer-focused proposals**—to sell when the salesperson can't be there

- **The Objective Based Selling diagram**—a memory tool

The open-ended questions create momentum in the sales process illustrated by the **Objective Based Selling** model, moving the customer toward a commitment to the salesperson's proposal, an order, or contract.

To be successful, the salesperson must meet both the business and personal objectives of the customer, working on both the business and personal sides of the **Objective Based Selling** loop simultaneously.

If the salesperson has the best proposal but has not established a strong, trusting, personal, professional relationship, the customer will find a reason and a way to buy from someone else. If that relationship exists, the customer will help the salesperson modify his proposal to be more responsive.

In the **Objective Based Selling** model, closing the order is really a customer commitment as a culmination of the **Objective Based Selling** process, rather than a decision triggered by a closing technique of the salesperson. The decision to commit to the salesperson is, in fact, often made when the salesperson is not present.

Nationally known sales and marketing consultant and speaker, Don Beveridge, must have had the material handling industry in mind when he said (paraphrasing):

> *You will never again, for any significant period of time, have the advantage of product or price.*

In that case, the sales skills and techniques of the **Objective Based Selling** model are essential to helping customers understand they can meet their business and personal objectives by acting on the salesperson's customer-focused proposal.

What's in your sales model?

OBJECTIVE BASED SELLING

Appendix 1

The Open-Ended Questions

One of the four keys to **Objective Based Selling** is open-ended questions, asked of customer decision influencers, by material handling salespeople. These questions are used to:

- Uncover business and personal objectives
- Build customer rapport
- Create motion in the sales process

This book has contained, in bold face, over 100 specific questions for material handling salespeople to use. This appendix has many of the most effective questions organized under the elements of the **Objective Based Selling** diagram. There are also some questions listed here which were not included in the book text, due to space constraints there.

In some cases, the notation (_____) will appear instead of words. This indicates specific words should be inserted, depending on the situation. For example:

"What is your biggest (_____) issue?" might be

"What is your biggest <u>forklift</u> issue?" or
"What is your biggest <u>space</u> issue?" or

"What is your biggest <u>loading dock</u> issue?"
"What is your biggest <u>safety</u> issue?"
"What is your biggest <u>caster</u> issue?" or
"What is your biggest <u>order picking</u> issue?"

The order of the questions under each diagram heading is not meant to be sequential. Further, all questions under each heading are not necessarily to be asked in every customer meeting at that stage of the process. Which questions to ask in what sequence, and when, are elements of the individual salesperson's skill, their "art" of selling. The purpose of this list is to act as a reminder and organizer of the most effective questions in the **Objective Based Selling** model.

"I kept six honest serving men;
They taught me all I knew.
*Their names were **what**, and **why**, and **when**,*
*And **how** and **where**, and **who**."*
—Rudyard Kipling

Open-ended questions of Objective Based Selling

Prospecting

"Who should I talk with about (material handling) (forklift) (loading dock) (warehousing) (____) operations?"

"When can I schedule an appointment to learn more about your (____) operations and how we might help?"

"If you could change one thing about your (____) operations, what would it be?"

"What is your biggest (____) issue?"

"What one thing in your (material handling) (____) operations, if you could do it, would positively change the way you do business?"

"What is your worst (forklift) (order picking) (____) issue?"

"Who has responsibility for (safety) (material handling operations) (training) (_____)?"

"Who else in your operations do you recommend I talk with?"

"If you were me, who else would you want to talk with in your company?"

"Who in your company gets involved in (_____)?"

"Who are your biggest suppliers? customers? competitors?"

"Who else do you know in this (area) (industry) (_____) that I ought to know?"

"Who do you know that could use my help?"

"What professional associations do you belong to?"

"Who can you refer me to that might have similar issues? responsibilities?"

"What capital projects are budgeted for this year? For next year?"

"What uncommon or special circumstances cause you problems?"

"How satisfied are you with your current supplier of (_____)?"

"What dissatisfaction do you have with your current supplier of (_____)?"

"What triggered your interest in (_____)?"

"What application did you have in mind when you requested that information?"

"What brought you down to the show?"

"What (projects) (capital budget items) (improvements) is your company focused on this year?"

"What's changing in your (area of responsibility) (operations) (company)?"

"What are your key priorities this year?"

"I understand you don't have a current need; who do you know that might have a current need?"

Determine business objectives

"What are your objectives for the (_____) project?"

"What are you trying to accomplish?"

"What triggers this project?"

"What's the most important thing about this project?"

"What is the time frame for this project?"

"Why is that time frame critical?"

"Why is that deadline important?"

"What are the financial considerations in this project?"

"What is the budget?"

"How was that budget established?"

"Who gets involved in setting the budget?"

"What do you do when project estimates are over budget?"

"How do you plan to finance this purchase?"

"Describe for me the end result you'd like to see from this project."

"How will you define completion?"

Identify decision influencers

"What is your area of responsibility?"

"Who, besides yourself, is involved in this project?"

"What is your role in the decision-making process?"

"Who else should I be talking with?"

"Who will be involved in the (_____) aspect of the project?"

"Who does that person report to?"

"Who has overall responsibility?"

"What is your role in the process? project?"

"Who has the most say in this project?"

"Who is against the project?"

"Who is the strongest advocate for this project?"

Establish physical parameters

"What are the key parameters of your operation?"

"Who should we ask to be sure?"

"Who can tell us?"

"How can we verify that?"

"How were these determined?"

"Which are the most critical parameters?"

"What parameter range would cover 90 percent of your requirements?"

"What did I forget to ask that we need to consider?"

"What did my competitors ask that I didn't?"

Prioritize and negotiate objectives

"How would you prioritize these objectives?"

"Which are the most (important) (critical) objectives?"

"How can we get all the decision influencers together for a scrum meeting to review objectives, parameters and concerns, to make sure we're all on the same page?"

Develop alternatives

"What would your company's interest be in (saving 20 percent of the cost with fewer features) (_____)?"

"What alternatives would you like to consider?"

"What would your reaction be to (_____)?"

"How about an alternative approach that would (_____)?"

"What if we (_____)?"

"Think through this with me: (_____)."

Review proposal with customers

"Who should be there when we review the proposal?"

"How can we get appropriate decision influencers together to review the proposal?"

"What's changed since we last talked?"

"What would you like me to focus on in this proposal review?"

"What is the most important thing about this project?"

"What are you most concerned with in this project?"

"What happens to this document after this proposal review?"

"What will your criteria be to select a supplier?"

"If all the prices were the same, what would your criteria be?"

"What will your criteria be to decide whether to go forward on this project?"

"Who in your organization is opposed to this project?"

"Who is the champion for this project?"

"What is your time frame for a decision?"

"What is your reaction to this proposal?"

"How does this proposal address your objectives?"

"What does this proposal leave unanswered?"

"Which of these alternatives do you favor?"

"What did you see in competitive proposals that you liked? That surprised you?"

"What should we be doing to follow up with you following this meeting?"

"What is your time frame to move this project forward?"

"What's our next step?"

Provide proposal support

"What elements of our proposal would you like back-up information for?"

"What references would be most meaningful to you?"

"What are your objectives for a product trial? demonstration?"

"Who should be there?"

"How will we know it's been a successful demonstration?"

"What should we focus on in this demonstration?"

"What's your reaction to the demonstration?"

"What are your objectives for a site visit?"

"Who should go with us?"

"What more do you need to know about our company? Capabilities? Proposal? Equipment? Performance record?"

Modify proposal

"What changes were a catalyst for this modification?"

"Why are we modifying the proposal?"

"What's the time frame for a decision following proposal modification?"

"What triggers this modification?"

"Who triggered this modification request?"

Identify personal objectives (and establish rapport), Build personal, professional relationships

"What is your area of responsibility?"

"How long have you had this responsibility?"

"What are your key responsibilities?"

"How long have you been with this company?"

"What do you like about working here?"

"What don't you like about working here?"

"What's the best thing about this (company) (responsibility)?"

"Who were you with before you came here?"

"What kinds of things do you do when you aren't working?"

"What do you do for fun?"

"What professional organizations are you involved with?"

"Tell me about (some item, plaque, picture in their office)."

"Tell me about yourself."

"What are your career objectives?"

"How would you like me to communicate with you?"

"What's the best way to communicate with you?"

Obtain commitment (close the order!)

"What can we do to move forward with this project?"

"What's our time frame to proceed?"

"Which alternative should we proceed with?"

"What stands between us doing business together?"

"What should we be doing next?"

"Where does this proposal come up short?"

Follow through

"How has this (project) (equipment) (service) (process) (implementation) met your expectations?"

"How have we met your expectations on this project?"

"Where have we come up short of your expectations?"

"What issues are left unresolved?"

"What did you like best about the way we handled this project?"

"What should we have done differently?"

"How can we define completion of the project?"

"If you were doing the project again, what would you do differently?"

"Who else in your organization is considering capital projects? Similar projects?"

"If you were me, who would you want to show this project to?"

"What other organizations do you know that might benefit from this type project?"

"Based on your satisfaction with this project or purchase, would you write a brief letter of recommendation, or satisfaction with our work?"

"What is our next project?"

Customer-Focused Proposal Templates

The "long form" proposal template (page 193)

This multi-page template is to be used for the larger, more complex, significant projects—projects justifying this effort. It is followed by two alternative forms: a three-page "proposal sandwich" format, and a single-page format.

All of these are basic starting points, to be modified to meet the needs of given sales situations.

The basic feature of all of these templates is they are customer-focused, instead of product-focused. Their purpose is to sell: even when the salesperson is not present. These templates start with the customer, and build a bridge to the recommended product or service, followed by proposal support.

The "proposal sandwich" proposal template (page 198)

This is for use with projects that are smaller or simpler but still of significant size, scope, or importance.

"Short form" single-page proposal template (page 201)

Some proposals are small or simple enough that a single page (plus possibly a specification sheet or similar item) is sufficient. Following is a suggested template for those situations. This should be formatted to fit the salesperson's company's proposal form.

"Long form" proposal template

Inside Address
Attn:
Re: (_____) Project

Dear NAME:

Thank you for the opportunity to work with you on your (_____) **Project**.

Accompanying this letter is our proposal for (name of equipment or service being proposed). This proposal includes:

- ■ (_____) Company objectives
- ■ Parameters
- ■ Recommendations and Benefits *(including an alternative)*
- ■ Product literature and/or drawings, pictures
- ■ Bill of material
- ■ Investment proposal *(includes commercial terms)*
- ■ References *(list of customers for whom you have provided similar equipment, or for whom you've done similar projects or provided similar services)*
- ■ Proposal support: Why buy from us!

This proposal is based on information provided by your company, and on examinations of your operations conducted by (_____) and me.

Again, thank you for this opportunity. We look forward to working with you in the implementation of this project.

Very truly yours,

Salesperson
Title

P.S. As appropriate, we can arrange a site visit to (_____)
(or alternative P.S. to help sell the project—they always read the P.S.!)

(_____) Company Objectives

As we understand them,

the (_____) Company objectives for

the (_____) Project, in order of priority, are:

■ _____

■ _____

■ _____

■ _____

Key parameters
in the (_____) Project

Based on information provided by (_____) Company

and our examination of your operations,

we understand key parameters in this project to be:

■ _____

■ _____

■ _____

■ _____

Recommendations

To assist (_____) Company achieve its objectives in the (_____) Project, we recommend the following:

(Include here a brief description of models, services, equipment to be provided. This can include drawings, literature, pictures, descriptions. It should not include detailed specifications; these will be included in the bill of material and investment proposal portions of the proposal. Include alternatives being recommended, with clear explanation of differences between them, as related to customer objectives and parameters.)

Benefits of these recommendations in meeting (_____) Company objectives

(This should be a list of bullet points as to why this is the right equipment and proposal for them, and perhaps a list of a few key features as they apply to this operation and this customer's objectives. A rate of return or payback calculation can be included here if these are part of this company's decision criteria. In this section, you are literally telling the customer why your recommendation and proposal meets their business objectives.)

Bill of material
Investment

Quantity

Specifications

Investment dollars

Terms of payment, taxes, freight arrangements, and other commercial terms

Delivery information (if appropriate)

(Clearly delineate differences in alternatives being recommended for consideration, including investment dollars. If more than two alternatives, consider spreadsheet presentation.)

Proposal support

(Always include references of other users of similar equipment or services, using a statement similar to that below.)

References

Following is a partial list of users for whom we have (provided similar equipment) (implemented similar projects). Complete contact information can be provided.

- (_____)
- (_____)
- (_____)

(List company names and brief descriptions of what was provided, objectives accomplished.)

Other proposal support can include:

- Pictures
- Reverse time lines
- Case histories
- Insurance certificates
- List of team members, project manager, key suppliers...

Proposal Sandwich Template

Inside Address
Attn:
Re: (_____) Project

Dear NAME:

Thank you for the opportunity to work with you on your (_____)
Project.

Project objectives, as I understand them, are:
- ■ (_____)
- ■ (_____)
- ■ (_____)
- ■ (_____)

Key parameters include:
- ■ (_____)
- ■ (_____)
- ■ (_____)
- ■ (_____)

Included here is our proposal for (_____) to help you meet these
objectives.

I look forward to working with you on this project.

Sincerely,
Salesperson
Title

P.S. *(Add something to help sell the proposal—they always read
the P.S.)*

Bill of material Investment

Quantity

Specifications

Investment dollars

Terms of payment, taxes, freight arrangements, and other commercial terms

Delivery information (if appropriate)

(Clearly delineate differences in alternatives being recommended for consideration, including investment dollars. If more than two alternatives, consider spreadsheet presentation.)

Benefits of this proposal
relative to your objectives

- (_____)

- (_____)

- (_____)

- (_____)

References

Following is a partial list of users for whom we have (provided similar equipment) (implemented similar projects). Complete contact information can be provided.

- (_____)

- (_____)

- (_____)

"Short form" single page template

Customer's Address
Attn:
Re: (_____) Project

Thank you for the opportunity to submit this proposal for your
(_____) **Project.**

Objectives, as I understand them, for this project are:
- ■ (_____)
- ■ (_____)

Key parameters include:
- ■ (_____)
- ■ (_____)
- ■ (_____)

To meet these objectives, I propose (_____), model (_____)
equipped with:
- ■ (_____)
- ■ (_____)
- ■ (_____)

See enclosed brochure for additional details.

Investment: $(_____)
Terms and conditions on other side of form

Other users of this product include:
- ■ (_____)
- ■ (_____)
- ■ (_____)

Salesperson
Title

Giving the Customer Homework: Surveys

There is a tool which can combine several elements of the **Objective Based Selling** model:

- Prospecting
- Determining objectives
- Identifying decision influencers
- Establishing parameters
- Prioritizing objectives
- Developing alternatives

This tool is: Give the customer homework! Ask the customer to fill out a material handling survey.

A material handling survey is a series of questions about the customer's material handling operations, which help define their current situation; parameters; areas of needed improvement; pertinent data; objectives of planned material handling projects; decision influencers; concerns; and impacting factors. These questions are organized into a paper or electronic form, designed to both educate the customer as to key factors they should be looking at, and eliciting information from them.

Two examples are provided at the end of this appendix:

■ **Client Overview Survey:** Data for material handling, space, and operation analysis

This survey is designed for general material handling and storage activities in distribution operations.

■ **Forklift Fleet Survey**

This survey is designed specifically to deal with issues involved with a forklift fleet used in manufacturing, distribution, yard, or other high use operations.

The reader is encouraged to review these forms before further reading in this appendix.

These forms are starting points. With word processing capability, they can and should be modified for specific situations and customers. Other material handling and facility specialty surveys often designed and used are:

■ Loading dock safety surveys

■ Facility door maintenance / Operational surveys

■ Caster application surveys

■ Order picking surveys

■ Space utilization surveys

■ Forklift battery condition / Maintenance surveys

■ Energy conservation surveys

■ Loading dock operational surveys

■ Forklift operator training practices / Compliance surveys

Actually, a material handling survey can be quickly designed from most material handling situations.

When are they appropriate to use?

Material handling salespeople are often called upon for their specialized expertise. Customers often to say things like:

"What can we do to make these operations better?"
"How can we reduce our (_____) operating costs?"
"Where can we find more space?"
"How can we modernize this area?"

The more vague the request, the more work they are asking the salesperson to do—without assurance of payoff with an order—the less likely they will even do anything.

In a sense they are asking the material handling salesperson to act as a consultant, without committing to pay a consulting fee.

In other situations, a customer obviously needs help in an area of his operations, but is simply unsure where to start, or doesn't even recognize the issues.

By presenting the customer with a survey form, the salesperson is indicating a level of professionalism above many others in the field. Just knowing what questions to put on the survey form is an indication of a level of expertise.

When the salesperson then asks the customer to fill out the survey, so he can use the information provided to make recommendations for improvements in the customer's operations, the salesperson is:

■ Asking the customer how serious he really is about improving his operations. If he won't get a survey form filled out—won't give the salesperson information to work with—how serious can he really be?

■ Making a trial close. If the customer agrees to fill out the survey, it's a commitment to the salesperson—a commitment competing salespeople probably won't get.

■ Potentially getting access to information about the customer's operations which will give him the opportunity to offer several suggestions—which hopefully will result in projects to purchase equipment or services from the salesperson. The salesperson is making something happen.

■ Avoiding doing a lot of work himself for the customer without any commitment from the customer.

The survey approach is most effective when made to a "higher up" decision influencer. This is because this person will see the benefit of getting this information together, but most likely won't have to do it himself. He can assign it to someone of lesser rank in the organization, someone who might see it as an opportunity to prove himself and get some recognition within the organization.

When the customer calls to say the survey is filled out—the homework is done—it is a sign to the salesperson that the customer is getting more interested in the project. It is time to check their homework!!!

Client Overview Survey

Data for
Material Handling,
Space and Operation Analysis

Brief Mission Description:

_____

_____

_____

_____

Completed by: _____

– 2 –

Operating Departments:

NAME   LOCATION

_____   _____

_____   _____

_____   _____

_____   _____

What are the primary measures of performance for this company (Division)?

_____

_____

_____

Projected Growth/Addition of Business:

One Year: _____

                       _____

Two Years: _____

                       _____

Three Years: _____

                       _____

Four and Five Years: _____

                       _____

– 3 –

If you could change one thing about your material handling operations, what would it be?

_____

_____

_____

_____

_____

What would you consider your primary objectives of any material handling or storage changes in descending order of priority?

A. _____

_____

_____

_____

B. _____

_____

_____

_____

C. _____

_____

_____

_____

Where do you need more space?

_____

_____

_____

– 4 –

What is perceived as the biggest factor in coping with projected company growth? _____

_____

_____

What is perceived as the greatest overall "bottleneck" in material handling and storage? _____

_____

_____

_____

As you consider your company goals and objectives, which of the following potential problems might you be considering?

____ Expansion   ____ Renovation   ____ Relocation   ____ New Construction

What problems concern you most as you relate to your existing storage and production areas? (Check applicable problems and assign priority.)

	PRIORITY					PRIORITY		
	1	2	3			1	2	3
( ) Space	( )	( )	( )	( ) Travel Time (Internal)		( )	( )	( )
( ) Retrieval Time	( )	( )	( )	( ) Travel Time (External)		( )	( )	( )
( ) Inventory Control	( )	( )	( )	( ) Labor Force		( )	( )	( )
( ) Down Time	( )	( )	( )	( ) Housekeeping		( )	( )	( )
( ) _____	( )	( )	( )	( ) _____		( )	( )	( )

If you were going to make changes in your facility or production/ storage layouts, which area would you attack first? _____

_____

_____

Why? _____

_____

_____

Financial Justifications

What are the formal criteria used for financial justification of capital equipment acquisition or construction?

_____

_____

_____

_____

Over what dollar figure must capital equipment acquisitions be submitted for formal approval?

_____

When may capital equipment requests be submitted?

_____

_____

How long does the approval process take?

_____

Who, besides yourself, is involved in this approval process?

    NAME   TITLE

_____ _____

_____ _____

_____ _____

_____ _____

Please attach any capital equipment request forms.

Financial Justifications
Page 2

What is the dollar figure per square foot of facility space saved allowed for use in capital equipment justification? _____

What is the estimated cost of new construction or leased space (per square foot) in this area for your type of facility? _____

What is the average labor cost allowed for labor savings in this facility (by department, if appropriate)?

DEPARTMENT	AVG. WAGE RATE	PLUS	FRINGES	TOTAL COST
_____	_____		_____	_____
_____	_____		_____	_____
_____	_____		_____	_____
_____	_____		_____	_____

Other possible considerations? (Please check those which may provide further justifications, or which need focus for improvement.)

( ) 1. Safety
( ) 2. Reduced stockouts
( ) 3. Cleanliness
( ) 4. Product obsolescence
( ) 5. Mobility and rearrangement
( ) 6. Flexibility to change, expand and grow
( ) 7. Security from pilferage
( ) 8. Cosmetic value as it relates to morale and cleanliness
( ) 9. Better supervision without total visibility
( ) 10. Ease and simplicity when taking inventory
( ) 11. Improved manufacturing – output and support

Further information on those items checked: _____

_____

_____

_____

Forklift Fleet Survey

FOR

_____

_____
Date

Completed by: _____

– 2 –

How many lift trucks in your fleet? _____

Do you own or lease your forklifts? _____

Please provide an inventory list of your forklifts, including the following information:

        Your Unit Numbers   Model
        Make   Serial Numbers
        Year of Manufacture   Special Equipment
        Department Assignment   Hour Meter Readings
        Maintenance Cost Summaries Per Unit

Any comments regarding condition or operation of specific units.

What is your forklifts' primary power (fuel) source:
        ____ Gas   ____ LPG   ____ Diesel   ____ Electric

If you use electric forklifts, please provide a list of motive power batteries.

How do you determine if you have the right number of forklifts? _____

_____

_____

OVERALL OPERATION:

What are your biggest concerns about your forklift fleet and its operation?

_____   _____

_____   _____

What are your biggest concerns about your forklifts?

Maintenance? _____

_____

Safety?_____

_____

Parts? _____

_____

– 3 –

Operators? _____

_____

Refueling or battery charging? _____

_____

Tires?_____

What specifications would you change if you acquired new forklifts?

_____

_____

What key things would you like to change, manage, or control better about your forklift fleet?

_____

_____

_____

RENT

How often do you rent forklifts? _____

What triggers this? _____

_____

What size/style trucks do you usually rent?_____

_____

What is typical length of rental?_____

MAINTENANCE

Describe the current maintenance program for your forklifts: _____

_____

– 4 –

Who does it? _____

What do you like/not like about this program? _____

_____

RECORD KEEPING

Describe your system for collecting maintenance and cost history for your forklifts. _____

_____

How is this reported? With what frequency? To whom?

_____

_____

Please provide any available maintenance/cost history for our review and analysis.

PARTS

Where do you get your forklift parts? _____

_____

What would you change about this? _____

_____

Do you keep a forklift parts inventory?_____

How big is it? _____

Please supply a history of forklift parts purchased in the past year.

OPERATORS AND SAFETY

Describe your Forklift Operator Training program: _____

_____

– 5 –

What operator issues do you have with your forklifts? _____

_____

How do you control operator abuse to your forklifts? _____

_____

What safety concerns do you have about your forklift operations? _____

_____

MANAGEMENT

Who, besides yourself, gets involved in management of your forklift fleet?

RESPONSIBILITY AREA	NAME	TITLE
Overall Operation	_____	_____
Maintenance	_____	_____
Parts	_____	_____
Fleet Replacement	_____	_____
Purchasing	_____	_____

What decision criteria do you use to determine when to replace forklifts?

_____

What plans do you have to replace or add forklifts? _____

_____

What is your timeframe? _____

What are your primary criteria when selecting new forklifts?

• _____

• _____

• _____

• _____

Acknowledgments

We all learn from others. Much of what I synthesized into **Objective Based Selling** was drawn from observations and experiences working with other professionals throughout my career.

And, in putting this book together, I relied on help and encouragement from several colleagues, friends, and family.

I acknowledge and thank:

- Ray Windas, Jr. of General Motors Assembly Division, North Tarrytown, New York for challenging me to do the tough stuff, and setting an example by doing it himself.

- Warner Frazier of Allis Chalmers Industrial Truck Division for hiring me into this challenging, rewarding industry.

- Doug Vokes of Allis Chalmers Industrial Truck Division for teaching me my first sales question.

- Jack Patten and Robert Patten of Materials Handling Equipment Company for offering me the opportunity to work in their company, and providing me amazing opportunities. They also first suggested I write a book on **Objective Based Selling**.

- Ralph Norton of Stanley Vidmar Company for teaching me it's not about the catalogues.

- Tom Flanagan of Allis Chalmers Industrial Truck Division for opening my eyes to the power of relationships in selling.

- Don Bray, Wayne Halligan, Harry Neumann Jr, Vicki Rains, and Richard Rankin of Materials Handling Equipment Company for teaching me a lot about selling—each in his and her own way.

- Rite Hite Corporation for introducing me to scrum meetings.

- Ron Dahlin and Warren Gandall, then of Eagle Material Handling, for giving the first opportunity for **Objective Based Selling** outside of Materials Handling Equipment Company.

- Dan Reilly of the Material Handling Equipment Distributors Association (MHEDA) for offering me early industry exposure.

- Liz Richards of MHEDA for encouraging me and offering opportunities in this and many other projects. She also gave me many valuable suggestions.

- Joe Verzino and Bill Rivett of Liftech Equipment Companies; George Sefer of Atlas Toyota Material Handling; Loren and Scott Swakow of Scott Lift Truck for offering suggestions on the text and project. Thanks also to Mike Dubbs of Storage Equipment Inc. and Dennis Hensen of Siggins Company for their comments about **Objective Based Selling.**

- Judy Joseph of Paros Press for guiding the publishing process; Susan Remkus of Copywriting etc., my editor; and Scott Johnson of Sputnik Design Works, who designed the graphics of the book and cover.

- Ron Conrad of Materials Handling Equipment Company for his integrity, work, and support.

- Robert Hochstadt of GHP Horwath for his friendship and professional advice over many years.

- My wife, Jane Costain, for giving up her dining room table for months—and supporting me in completing this project.

I have, of course, learned from many others throughout my career. I hope I have helped others along the way—and that this book helps material handling industry salespeople sell more... at higher gross margins!

Gary T. Moore
September 2007

To order additional copies of

Objective Based Selling™

please visit our website at
www.trentpress.com

You can also contact **Gary Moore** at gmoore@trentpress.com regarding other **Objective Based Selling**™ training opportunities—

- One day **Objective Based Selling** seminars for your sales force
- Sales Managers' workshops, including how to implement **Objective Based Selling** as your distributorship's sales model
- **Objective Based Selling** presentations for material handling supplier meetings for their distributors
- Material handling industry seminars and workshops on
  - Selling more effectively through material handling distributors
  - Effective marketing in the material handling industry
  - Business development for distributors and suppliers
  - Customer service excellence

Gary also consults with distributors and suppliers on these and other distribution related topics.